Studies in Big Data

Volume 47

Series editor

Janusz Kacprzyk, Polish Academy of Sciences, Warsaw, Poland
e-mail: kacprzyk@ibspan.waw.pl

The series "Studies in Big Data" (SBD) publishes new developments and advances in the various areas of Big Data- quickly and with a high quality. The intent is to cover the theory, research, development, and applications of Big Data, as embedded in the fields of engineering, computer science, physics, economics and life sciences. The books of the series refer to the analysis and understanding of large, complex, and/or distributed data sets generated from recent digital sources coming from sensors or other physical instruments as well as simulations, crowd sourcing, social networks or other internet transactions, such as emails or video click streams and others. The series contains monographs, lecture notes and edited volumes in Big Data spanning the areas of computational intelligence including neural networks, evolutionary computation, soft computing, fuzzy systems, as well as artificial intelligence, data mining, modern statistics and operations research, as well as self-organizing systems. Of particular value to both the contributors and the readership are the short publication timeframe and the world-wide distribution, which enable both wide and rapid dissemination of research output.

More information about this series at http://www.springer.com/series/11970

N. Jeyanthi · Ajith Abraham
Hamid Mcheick
Editors

Ubiquitous Computing and Computing Security of IoT

 Springer

Editors
N. Jeyanthi
School of Information Technology
  and Engineering
VIT University
Vellore, Tamil Nadu, India

Hamid Mcheick
Université du Québec à Chicoutimi
Chicoutimi, QC, Canada

Ajith Abraham
Scientific Network for Innovation
  and Research Excellence
Machine Intelligence Research Labs
  (Mir Labs)
Auburn, WA, USA

ISSN 2197-6503 ISSN 2197-6511 (electronic)
Studies in Big Data
ISBN 978-3-030-13186-9 ISBN 978-3-030-01566-4 (eBook)
https://doi.org/10.1007/978-3-030-01566-4

This Springer imprint is published by the registered company Springer Nature Switzerland AG
The registered company address is: Gewerbestrasse 11, 6330 Cham, Switzerland

Contents

Security Protocols for IoT

J. Cynthia, H. Parveen Sultana, M. N. Saroja and J. Senthil

Abstract The Internet of Things (IoT), is a network of devices that are uniquely identified and has embedded software required to communicate the transient states and data that are usually used to trigger an actuator. The edge networking devices and protocols are used to communicate with a cloud server that processes and aggregates the big data arriving from various devices, performs analytics and aids in business decisions. IoT has become an integral part of today's industrial, agriculture, healthcare and smart city revolution. Securing all entities involved in an IoT network is vital as it involves pervasive data collection and dissemination. Current IoT protocols work with IP protocols as backbone, but they are specially designed to operate in multiple layers and provide security at various layers. This chapter focuses on IoT protocols that deals with securing an IoT network. The major challenges in securing an IoT network is lack of standardization at manufacturing level which exposes the hardware, software and the data to various threats and attacks. The IoT protocols have to deal with security breaches at the site of the cloud service provider and the security issues pertaining to data privacy, authentication, authorization and trust management in a distributed heterogeneous environment. This chapter also elaborates on various security attacks and the solutions offered by IoT protocols.

Keywords IoT security · IoT architecture · IoT protocols · IoT threats
IoT attacks · Heterogeous

1 IoT Introduction and Security Overview

There are over billions of IoT devices, business process and systems with an IoT element in it. This dormant data available in the eco system must be trapped and

---

J. Cynthia · M. N. Saroja · J. Senthil
Kumaraguru College of Technology, Coimbatore, India

H. Parveen Sultana (✉)
VIT University, Vellore, Tamil Nadu, India
e-mail: hparveensultana@vit.ac.in

© Springer Nature Switzerland AG 2019
N. Jeyanthi et al. (eds.), *Ubiquitous Computing and Computing Security of IoT*,
Studies in Big Data 47, https://doi.org/10.1007/978-3-030-01566-4_1

analyzed to see if any functional information that is of benefit to a customer or business could be retrieved. Essential elements of IoT are people, things, data and process. IoT systems aims at networking these elements that communicates with each other through wired or wireless medium. IoT devices are grouped as sensors that collect data, Actuators that effect actions and gateways that act as interface for communication and automation. In an IoT framework, data is gathered from sensors, processed by microcontrollers such as Raspberry Pi or Arduino, stored in a cloud database and data analytics from big data gathered is performed using any tool or languages such as python or java. IoT is designed to strengthen communication across Device to Device (D2D), Human to Device (H2D), Human to Human (H2H) and Device to Human (D2H).

IoT has led to numerous autonomous applications in the area of health care, business solutions, smart city, home automation, industry automation and intelligent transport system. The success of IoT lies in distributed data gathering, aggregation, processing and analytics that can be performed from any location and is usually done as a cloud service. IoT system evolves with flow of data from the sensor from where it is acquired to the service that processes and performs analytics on the data acquired to the customer or business that makes use of the analytics information.

With prevalent presence of IoT, security risks are in rise. Making data available anywhere makes it vulnerable to security threats and attacks. This chapter deals with major issues, challenges and solutions for providing IoT security. A single compromised entity in an IoT network makes other entities vulnerable. Since IoT is a collection of devices or sensors networked together to a cloud in order to provide information service, all security threats that are applicable for Wireless Sensor Networks (WSN), internet and cloud are pertinent to IoT networks. IoT opens up tremendous opportunity for business with the associated risk. Absence of strong authentication of IoT devices, encryption of IoT data, key management, etc., makes an IoT network vulnerable to external attacks and threats.

2 IoT Security Requirements

Security must be addressed throughout the lifecycle of an IoT device. Shipley [1] and Jing et al. [2] lists security requirements to be checked at various stages of the life cycle in order to alleviate an IoT attack. IoT security requirements are listed below,

- Cryptographic Algorithms—Symmetric algorithms are light weight compared to asymmetric algorithms and hence were recommended for securing data transmission. However, they have problems in key exchange, confidentiality, digital signature and message authentication. Hence public key algorithms were recommended as they were able to provide key management, node authentication, scalability and security.
- Key Management Techniques—Key management is an important security feature in IoT. Light weight secure key distribution is required for secure communication.

Key distribution schemes used in WSN are broadcast, group, node master and shared key distribution [2]. The focus on key management research is to reduce the complexity, power consumption and security.

- Secured routing algorithms—Traditional network routing protocols cannot be applied for IoT network. The routing protocol must ensure authenticity of routed information and eaves dropping must be avoided while communicating through wireless medium. Routing protocols should be secured to prevent attacks such as Dos, Worm hole, black hole and selective forwarding.
- Data Classification—The data floating in an IoT network could be either functional or connected to people or an enterprise. The degree of protection required for a data depends on the degree of sensitivity of the data. Data may be protected based on sensitivity classification [3]. Hence following recommendation is made for an IoT vendor,

  - To define a data classification scheme based on data sensitivity.
  - Identify all data and data groups in an IoT network and classify them.
  - Design a security feature that protects viewing and editing of data based on its classification level.

- Protecting devices at production time—The IoT devices may be protected at production time. Any interface used at production time must be removed before deployment. All ports to the IoT devices must have proper access control. Devices placed in exposed locations must have a tamper proof covering and shielding to avoid side channel attacks [3].
- Trusted and staged boot sequence—A trusted staged boot sequence will ensure security of an IoT device. However, the first sequence is vital and hence should be initiated by secured locked code. Use of secure module where the cryptographic algorithms and associated keys are stored are recommended. At every stage of boot code, it is recommended to check the trust worthiness of boot code, validity of hardware and completion of previous code.
- Secured operating system—An IoT operating system should have limited access rights and reduce the visibility of the system. The operating system should be designed so as to have only the components, packages and libraries required for running an IoT device. Throughout the lifetime of the deployed device the update must be provided. The ports, protocols and services that are not used are to be disabled. Have separate access rights for user and administrators to access the files and directories must be given. An encrypted file system is to be used.
- Application security—Security considerations must be an integral part of application development and should not be added separately. The application gateway should validate all gathered data before it is getting processed. All user accounts and passwords are to be relinquished. Credentials from application has to be separated into a secured storage. Any application errors should not reveal details about the underlying architecture. Use of secured software development life cycle procedure is recommended.
- Credential management—Credentials such as passwords, cryptographic keys and digital certificates of user and process that are used to access the data must be kept

in secured location that cannot be accessed by external entities. The passwords used for authenticating must be strong, encrypted and must have industry standard hash function. Two factor authentications may be used for access control. Unique digital certificate for each device is recommended and this certificate must be secured and updated at regular intervals.

- Encryption—Strongest and latest encryption is recommended for an IoT network, if it is affordable. The encryption standard should be in correlation with the sensitiveness of the data to be protected. Use of global keys is to be avoided. The private key of a device should never be shared. The encryption keys should be able to be replaced remotely. The encryption keys must be stored in trusted key modules.
- Network connections—The number of interfaces to an IoT device through which it gets connected to the external network must be kept as minimal. The device must be able to be accessed only through minimal port, interface and services. Secure protocols such as https and SFTP to protect connections are to be used. Receiver machine must be authenticated before sending any sensitive data.
- Software updating—Before any software updation, authentication of the source that authenticates, must be done with help of a verified certificate obtained from a authenticated certification authority. The software update packages must be signed.
- Secured event logging—The event logging should be protected from hackers, from being modified or deleted. The event logs are normally stored in a centralized log pool away from the IoT device and hence must be transmitted though separate channels. The logs must be periodically analyzed to detect any faults and immediate action is to be taken. The log files must be stored in separate partitions in file system. Access rights to the log file are to be restricted. No sensitive credentials such as passwords are to be stored in logs.

3   IoT Security Issues

The issues associated with security of IoT are not only the issues related with security of wireless medium, WSN and internet, but also access control, authentication and privacy issues associated with IoT.

- Low power embedded device—IoT devices have less computation power and storage capacity. It is often found embedded in a bigger hardware or wearable device where it is difficult to execute security algorithms that are normally heavy weight and expensive for a resource constrained device.
- Trust Management—Trust management is required for data authentication data gathering and dissipation phases for which strong cryptographic techniques or digital signatures are recommended [2].
- Heterogeneity—IoT is an integration of various heterogeneous networks and hence has its own compatibility and security issues. It is difficult to identify trusted nodes in a heterogeneous environment. Heterogeneity, identity management, privacy fault tolerance [3].

- Secured Access control—Secured access control is a major challenge in an IoT network. Usually the information in the cloud is accessed by various entities and process. Also the granularity level for accessing the same data differs for different retrievers. Therefore, defining access control policy and securing the access is one of the major challenges [3].
- Identity Management—It is required to uniquely identify an IoT device and provide both authentication and authorization for each of the device. Authentication ensures the validity of the data that flows through the device and authorization ensures secured access control. The entities in an IoT network may be added dynamically and hence identity management with authentication becomes even more difficult.
- Privacy—It is important to provide privacy for the billions of users in IoT networks. Anonymity of the user must be maintained. Access control list must be maintained by any service provider. Privacy must be given its due importance in the entire IoT life cycle.
- Trust Management—Trust management plays a vital role in communication across entities and between an entity and user. Reputation calculation is required to decide on a trusted entity. The collective view of a central entity helps in calculation of reputation of the remaining entities. The inconsistencies in the reputation value may be resolved by sharing the trust information from various central entities.
- Distributed IoT Network—A centralized or connected IoT network has separate data acquisition passive entities, which give the collected data to a centralized cloud service that does the job of aggregating, processing, analyzing and distributing. Moreover, the information flow to the central authority follows a hierarchical pattern. This has better centralized security control but once subjected to vulnerability, the entire system is compromised. In a distributed IoT network every entity is entitled to do the job of data collection, processing, analyzing and distributing information and hence is an attack vector.

  But however an attacker will be able to retrieve only the partial information from the attacked entity which may also be the vital information required. The edge intelligence at the service provider's end to query the information by a local user without intervention from any external entity has a potential vulnerability which should be controlled by providing strong authentication and authorization features [3].

4 IoT Security Challenges

Hossain et al. lists the challenges of IoT security based on limitations of hardware, software, network connections. The hardware limitations are, computational and energy constraint, memory constraint and tamper resistant packaging. Limitations on software are embedded software constraint and dynamic security patch. Limitations on network connections are mobility, scalability, multiplicity of devices and communication medium, multi protocol networking and network topology.

4.1 IoT Hardware

IoT hardware includes sensors, wearable devices, digital gadgets, microcontrollers like Arduino, Raspberry pi and embedded hardware. IoT hardware devices are present with the customers, embedded in some other device and may be used as a wearable device or may be present connected to the internet all time. Therefore, these devices are more vulnerable to security attacks and can be easily tampered with. Hardware device manufacturers are more concerned in design aspect of IoT devices rather than the security aspect. Hence the customers are exposed to more risk [4]. The reduced size and processing capability inhibits the security features of an IoT device [5]. Due to the prevalent presence of IoT hardware it is difficult to provide a software patch for security updates. Due to lack of standardization before manufacturing, also exposes the IoT hardware to security threats. IoT hardware are exposed to attacks to which all internet connected devices are exposed to such as DOS, and DDoS.

In order to protect the hardware, issues such as hardware lifecycle, software updates, access control and device authentication should be dealt with. Enterprises should take initiative to check the configuration of all IoT devices, perform vulnerability scan and check network connections [6]. Embedded system security is a major concern for growth of IoT. Various IoT consortiums are working on defining a framework to implement identity, device discovery, authentication and security controls in a consistent manner. Care should be taken to protect the private data present in hardware before they are discarded [7]. When choosing a hardware platform, the security concerns such as its unique identity and secured storage for encryption keys are to be verified. Evaluation to be done to check how difficult it is to change the credential stored in hardware.

4.2 IoT Software and Firmware

IoT software component includes the embedded software, operating systems used in IoT such as Android and Tiny OS, and cloud software such as Nimbis and Hadoop. Most of the IoT software deals with data gathering, integrating devices, application and process interface, and real time analytics. IoT devices connected to internet have operating system embedded as firmware. These operating systems are not designed with security concerns and hence are vulnerable to malware attacks. The embedded data in appliances, mobile phones and wearable devices with networking capability are more vulnerable to external attack. This is because they share the data with other connected devices and the embedded data lives for more period than the hardware themselves. The security aspect is neglected by the enterprise as the cost of hardware is much less than software and security upgrades. Improperly configured storage devices connected to network and are used from home are also major source of threat. There is huge volume of data generated from these devices. It is difficult to decide if the data has to be protected or not. Trojan horse or worms may be used to inject malicious code into software.

The most cost effective solution for protecting the embedded software is to monitor and secure the traffic at gateway [8]. The securities threats for the wearable devices used in health care and manufacturing sectors, can be minimized by disabling their bluetooth communication, geo fencing the communication, restrict communication and access control with external applications. Outdated operating system and software without a patch has to be avoided to ensure security.

4.3 Insecure Network Communication

Owing to the huge number of IoT devices connected to the network, tradition network security, identity and key management mechanisms are difficult to implement. Any device or process attached to an IP address or URL has an associated risk with it. It is difficult to bring the entire IoT device connected under the boundary of a controlled firewall, because an attacker may use a single compromised node to attack the entire network in a lateral manner. The monitoring and isolation of IoT devices involved to the private VLAN or network segment may reduce security threat [8]. Mesh network is suggested as a solution for connecting IoT devices, since it is Self-organizing, self-healing and scalable. Sudden increase in bandwidth requirement due to large volume of data generated from social networking sites and IoT will emulate the attack such as DoS. Wireless communication amongst IoT nodes subjects them to both active and passive attacks. A mesh network is formed by connecting wireless devices without any infrastructure. Meshing in IoT enables the IoT elements to communicate amongst themselves in absence of fixed infrastructure for communication. This is extremely useful in case of low power and low data rate applications in health care, industrial and home automation applications [9]. IoT network in an enterprise is subjected to vulnerability, if proper Enterprise Mobility Management (EMM) policy is not defined to mitigate the risk of vital corporate data leaked to the outside world.

4.4 Data Leaks from Cloud

Data is stored in a cloud with primary motive of sharing. Strongly authenticated sources in the Access Control List are expected to access the data. A service provider is responsible for any data leakage from cloud. A misconfigured cloud will lead to data leakage. External access to sensitive data and logs must be restricted. A hostile employee may gain access to any internal server and enterprises, outsource certain services with potential threat of data leak. Cloud environment demands continuous monitoring and intrusion detection. It requires monitoring and logging virtual machine logs and shared services. Intrusion detection and prevention systems are recommended for cloud in order to avoid data leakage.

4.5   Threats and Attack Vectors

The paper [7] indicates list of potential threats used by an IoT targeted attacker. He may use the MAC address to understand the target platform and reverse engineer the software, to find the encryption keys. The attack vectors are path used by a hacker to gain access to a secured system [10]. As the attack vectors available for malicious attackers are growing day by day because of the global connectivity and accessibility, fault tolerance must be provided. IoT data and meta data are potential attack vectors for any hacker. Following are some of threats in IoT,

- Denial of Service—A DoS attack in IoT is aimed at exhausting service providers' resources and network bandwidth. Channel jamming in wireless network is also a type of DoS attack. Since IoT devices are exposed to active attackers, it also leads to DoS type of attack.
- Eaves dropping—Passive attackers target the communication channel and eaves drop the data and extract the information. An active attacker may capture a node exposed to outside environment to gain access to the store data.
- Controlling IoT entity—An active attacker may gain control over an IoT entity through an attack path. This type of attack not only gains control of the data but also the services that are associated with the data.
- MQTT Attack—IoT servers that use Message Queuing Telemetry Transport(MQTT) on internet is subjected to attack because of unauthenticated and unencrypted communication. MQTT servers are also vulnerable to SQL injection and cross-side scripting. The MQTT servers used for firmware updates, may be used to update malicious code [11].
- Ransomware—IoT network are subjected to ransomware attack where they steal data from any interface gateway or cloud aggregator and claim money for the same. In a ransomware attack, an attacker usually gets hold of critical data as in hospitals that is required for day-to-day activity of an organization and demands money in some form to release the data. A ransomware attack in an IoT environment causes business loss.
- IoT Request Forgery—An attacker tries to target IoT devices connected to a corporate network rather than to crack several security layers.
- Wearable malware—The wearable devices acts as an attack vector of a mobile malware attack and allows authorized access to a connected IoT network, botnets have potential to attack IoT network as a group.
- Virtualization threats—The host machine running virtualization software could be attacked by code in virtual environment that simulates man in middle attack [12].

5   IoT Protocol Architecture

IoT protocol stack is not standardized as TCP/IP or OSI protocol suite. Most of the IoT security protocols are designed to operate in multiple layers to provide security.

The protocols used and security measures provided depends on whether a node is constrained or unconstrained [13].

Wireless Hart is a security protocol that operates in multiple layers using multiple keys and secures the traffic by encrypting payload and providing message authentication. Separate keys are used by network layer to authenticate end to end communication and data link layer to authenticate hop to hop communication.

LoRaWAN is the long range variant that provides secured bidirectional communication, mobility and localization services. It provides unique network key to ensure security in network layer, unique application key to ensure end to end security in application layer and also a device specific key.

As described in [14] illustrates the protocols operating in all 5 layers of TCP/IP protocol Stack, the IoT applications and associated services. Figure 1 illustrates the IoT architecture and protocol stack.

Physical Layer—This layer is data oriented and is responsible for collecting data from IoT devices. The issues to be considered in physical layer of an IoT network are power, bandwidth and energy consumption. The devices attached to this layer are susceptible to security challenges such as physical tampering of devices, eaves dropping and data altering. Cryptographic algorithms play a major role in physical layer security. Low power Wide Area Network (LPWAN) is used in IoT for transmission of small data over long range with battery efficiency. It uses modulation technique such as ultra-narrow band, narrow band and wide band. IoT connectivity technology

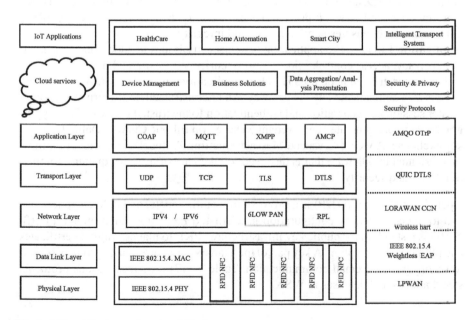

Fig. 1 IoT architecture and protocol Stack

may be chosen based on capacity of channel, QoS, reliability, range, battery life, security, cost and standard [15].

Data Link layer—Increasing the transmission power also increases the data rate in wireless communication. Any wireless communication protocol such as Bluetooth, Wi-Fi, Zig bee may be used. IEEE802.15.4 is used to provide link layer security. It protects MAC frames using symmetric key cryptographic techniques. This includes Zigbee, 6LoWPAN, Wireless-HART. Weightless is a standard used for exchanging data between base stations and several IoT devices in a secured fashion. EAP(Extensible Authentication Protocol) supports multiple authentication methods and runs independent of IP.

Network Layer—In network layer, security is usually provided by 6LoWPAN and IPSec protocols. A constrained node uses 6LoWPAN and an unconstrained device used IPv6 for addressing in IoT. 6LoWPAN is specifically designed to provide security in devices with low power and computing ability in WSN and internet. Therefore, cryptographic algorithms combine RSA and ECC techniques. 6LoWPAN must be accompanied by IDS techniques to monitor traffic for any malicious behavior [16]. CCN (Content concentric Networking) is a protocol used to deliver content as packets and has been designed to deal with scalability, mobility and security. IP Sec (Internet Protocol Security) is designed to provide authentication of sender data and encapsulating security payload to provide data encryption and sender authentication.

Transport Layer—QuiC [13] protocol provides multiplexed connections over UDP and provides security protection similar to TLS/SSL in order to reduce connection latency. DTLS (Datagram transport Layer) protocol offers communication privacy between client and server. It prevents eaves dropping, tampering and forgery. IPSec in transport layer ensures confidentiality and integrity.

Application Layer—The application layer security issues include user authentication, privacy, access control, middle ware security. A constrained node uses CoAP and an unconstrained node uses HTTP as application layer protocol. A constrained node is also authenticated by the gateway. An unconstrained node entrusts the job of master session key generation and authentication to the trusted gateway. The cryptographic keys are generated and exchanged based on Elliptic Curve—Deffie Hellman key exchange. AMQP (Advanced Message Queuing Protocol) is a protocol for message oriented middleware that is designed to take care of message queuing routing, reliability and security.

Open Trust Protocol (OTrP) is a protocol to install, update, and delete applications and to manage security configuration in a Trusted Execution Environment (TEE). X.509 is a standard for public key infrastructure (PKI) to manage digital certificates and public-key encryption. A key part of the Transport Layer Security protocol is used to secure web and email communication. Table 1 [17] describes various protocols stack, attacks and defenses in WSN.

Table 1 LLN protocol stack, threats and defense

Layer	Attack	Defense
Physical	Jamming	Channel surfing, spatial retreat, priority messages
	Radio interference	Delayed disclosure of keys
	Tampering	Tamper proofing, hiding
MAC	Collission	Error-correcting code
	Exhaustion	Rate limitation
	Unfairness	Small frames
Network	Sink-hole	Geo-routing protocol
	Worm-hole, black hole	Authorisation, monitoring redundancy
	Homing	Encryption
	Misdirection	Egress filtering, authorisation, monitoring
Transport	De-synchronisation	Authentication
	Flooding	Client puzzles
Application	Overwhelm	Rate-limiting
	Reprogram	Authentication

6 IoT Security Attacks

Internet of Things, the increasing need in our day-to-day life has more advantages. The important thing about IoT is, it makes the things beings intelligent by embedding sensors and actuators. By increasing the connectivity, it enables new services. On the other side, the amount of data generated by IoT is getting increased which results in security attacks.

Well, most of the people can think of

- Why Security is more important in IoT?
- What can a person do by attacking the device?
- Why is it important to consider the attack on device?
- Is it possible for my device to provide private data to intruders?

These are the questions will come to mind, when anyone think of security in IoT. Let me explain one by one.

Internet of things has a variety of sensors, wearable devices, mobile phones and home appliances. Most of the time, the devices are produced by the manufacturers who doesn't know about the security. Also, he is not a security expert too. When a user stores a private data such as mail passwords, bank details etc. in his mobile, he usually thinks that it is stored in his local memory. Actually it is stored in cloud storage. This will help the hacker to easily attack the data from the cloud and misuse it. In this way, the security in IoT is considered to be more important.

What may be the next question is usually, people will store data only in their mobile phones, then why there is a need to protect sensor devices and other home appliances?

The thing is, when a device, say a security camera connected to a home is attacked, the hacker can clearly know the possibility of robbing a house. This will invite unknown persons to home also.

Let me explain with another example. When a refrigerator which orders things needed for a smart home got hacked, he may order any number of things or he may generate a spam to randomly generate more things. In this situation, the user will either will lose money for things he has not ordered or will get irritated and switch off the device. The important thing is, the device which has been attacked is connected to our mobile phones and other devices also. So, when a simple device is attacked, the hacker can easily gain access to devices which contain secure data. This is the reason why security is considered to be important in IoT.

The next thing, we have to discuss is what are all the ways through which things/devices can be attacked. The various attacks that can be performed are firmware attack, data attack, telnet based attacks, denial of service attack. Let me explain them one by one.

6.1 Attacks on Firmware

Firmware is nothing but software used to control hardware devices. In the early 90 s itself, the firmware attack has been started. In general, firmware is stored in non-volatile memory. Hackers generally add some malicious code to this non-volatile memory and make it as a part of firmware and start controlling the device. Another reason why people prefer firmware attack is, they are harder to detect since they run before the antivirus program starts.

Hackers attack firmware for three main reasons [18]:

1. *Persistence*: Malwares can be cleared often using antivirus software, whereas firmware is not.
2. *Protection*: Mechanisms such as antivirus software's will not examine firmware so that it can be hidden and used for a long time.
3. *Authorization*: Being a part of firmware by adding malicious code, the user can get complete authorization for accessing the system.

The advantage with firmware attack is that the firmware software is obsolete often. Also, most of the people are unaware about updating device software. Of course, the manufacturer of a device is not an expert in security, which results in vulnerable device.

Do you think that the latest firmware will provide complete security? If you say yes, then it is your false sense about security. In reality, most of the devices which are manufactured recently are equipped with the operating system which is a decade ago. Also it was not maintained by Security professions which results in easier attacks.

If everything is negative, how to overcome this type of attack? Is it not necessary to update the software?

No, it is not so. The best suggestion which I could offer in this place is through updating the devices and keeps them up-to-date. Next thing is we, the consumers can demand the manufacturers to provide better security device. This can be done only when IT professionals, industry and security experts work together.

6.2 Attacks on Data

IoT enables more and more devices to be connected which results in more security vulnerabilities. The devices that are connected (as shown in Fig. 2) may include each and every object which we use in our day-to-day life.

We will store data from temperature to our sensitive data such as passwords. Is that protected over there?

No. We think that how a connected object can provide information. Consider a surveillance camera which records the data of a terrorist attack. If a person can hack it easily, he can change the records. Do you now understand the importance of data? Again, you may wonder that a surveillance camera is that much easier to attack? It may be protected, but the router connected to it or a sensor connected to it can easily be attacked. This way, the Internet of Things provide more options for the hackers

Fig. 2 Connected objects in IoT

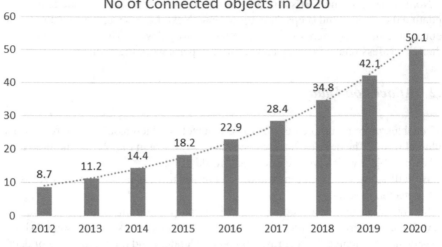

Fig. 3 Estimated no. of connected devices

to steal information. The estimated number of devices as per CISCO estimation may reach 50 billion by 2020 as shown in Fig. 3. If almost every device is vulnerable to attack, the world will not exist. If every data can be hacked by simple means, it doesn't make sense to be connected.

Let me explain with another example. You have stored your banking details (sensitive data) in your mobile phones. You think it is protected by passwords, fingerprints etc. so that it cannot be attacked. But when you are trying to control a fan (in Smart home—an IoT Application), using your mobile phone, it actually happens by means of a sensor. If sensor is hacked, then automatically malicious code is transferred to mobile devices which make it easier to attack. Now, do you think it is not possible? You cannot. The Internet of Things makes the things getting connected but also provide many security holes.

6.3 TELNET Based Attacks

This is an important topic in IoT. People will think Telnet is very old and what is there to be important in it. Hackers have changed the trend to use old techniques to attack new technologies. There is where the concept of Telnet comes. Telnet actually provides a gateway for attacking the internet of things. IBM Security has also released a research titled "Beware of Older Cyber Attacks" [19]. In that article, it is clearly highlighted that Telnet, a very old technique to access remote systems can be used as a key to gain access into unauthorized access.

Many embedded system applications leverage its remote access capabilities. If an attacker can find a open telnet port, then he can perform the following:

- Exploit any vulnerabilities associated with the device
- Gain unauthorized access to a device for stealing data
- Determine how the information is shared between devices
- Perform brute force attacks to gain passwords.

One example of this kind of attack is the Bricker Bot attack. The Bricker bot attack used Telnet Brute force attack to breach Victim's devices. Bricker Bot attack was designed to record the first attempted username and password. Through that, it will gain access to the devices connected to it. The attack can be blocked by disabling Telnet and changing the default passwords.

Another reason why telnet is important is most of the devices will be having default username and passwords. Even though people using the devices are instructed to change the passwords, it not clear that everybody does the same. The entire passwords are not changed. Such devices can be easily provided with remote access through Telnet and SSH.

6.4   DDOS Attack

Denial of Service Attack is another important attack in case of Internet of Things. Denial of Service attack generates more traffic to the server and overloads it which results in the service being rejected. If the DoS attack is performed with huge botnet, then it is called Distributed Denial of Service Attack. IoT botnets comprises of web camera's, TV, DVR, Setup boxes etc. to launch the DDoS Attack.

On 20 September 2016 [20, 21], "KrebsOnSecurity.com" [22] became the target of a massive DDoS attack that eventually knocked the site offline. The site was initially protected from this attack by Akamai, the website's digital security service provider. The company decided to withdraw its pro bono protection shield, since the magnitude of the attack (approximately 620 Gbps) was too vast to bear it without affecting other customers. Akamai's analysis indicated the use of a large botnet of compromised IoT devices. Upon Akamai's protection withdrawal, the website went offline until Google offered its DDoS attack mitigation service, Project Shield, to revive it.

OVH, a well-known Web hosting provider, was also a victim of an even more massive DDoS attack than the one that hit "Krebs on Security". According to a tweet from OVH founder Octave Klaba on 22 September 2016, a simultaneous DDoS attack of 990 Gbps (combined) was launched by a botnet consisting of more than 145,000 compromised IoT devices (IP cameras and DVRs). OVH reported that it withstood the attack.

Right after the DDoS attacks against "KrebsOnSecurity.com" and OVH, a user on a hacking forum released the source code of a malware dubbed "Mirai". The malware targets unprotected IoT devices and turns them into bots. The attacker is then able to launch a DDoS attack commanding all bots through a central command and control server as done in common botnets.

On 21 October 2016, the DNS provider Dyn, experienced a massive DDoS attack and initially claimed that the attack originated from tens of millions of IP addresses around the world. A later update from Dyn, noted that malicious endpoints were actually estimated to be around 100,000. The attack caused issues to certain users trying to reach popular websites such as Twitter, Amazon, Tumblr, Reddit, Spotify and Netflix throughout that day. According to Dyn's information on the Incident part of the attack involved IoT devices infected by the Mirai botnet. After several hours and several waves of attacks Dyn resolved the incident.

The main things about the massive IoT DDoS attacks are as follows:

1. Huge amounts of traffic at DNS servers made many websites to stop working.
2. Botnet is formed by large number of unsecured devices such as home routers and surveillance cameras.
3. Use of default passwords is one of the main reason for this vulnerability.

What can be done to secure from these things? I suggest you with the following solutions:

• Update IoT devices with security patches as soon as patches become available.
• Disable Universal Plug and Play (UPnP) on routers unless absolutely necessary.
• Purchase IoT devices from companies with a reputation for providing secure devices.

6.5   roBOT + NETwork (BOTNET)

IoT botnets are not new. A Botnet is a logical connection of compromised devices such as routers, smart phone or IoT devices. These compromised devices can be controlled and used for performing DDoS attacks. The objective of creating a botnet is to infect as many devices as possible. Generally, IoT botnets have been used to launch high-profile DDoS attacks against online gaming networks, to engage in DDoS extortion attempts, and to target organizations affiliated with the Rio Olympics.

Some of the notable botnet attacks are Zeus malware, Srizbi botnet, Gameover Zeus etc. The Zeus malware used a Trojan horse program to infect vulnerable devices and created a Zbot which can be used to harvest banking credentials and financial information. Srizbi botnet again used Trojan horse program. The Gameover Zeus botnet would generate domain names to serve as communication points for infected bots. An infected device would randomly select domains until it reached an active domain that was able to issue new commands

Mirai malware is designed to scan the internet for insecure connected devices, while also avoiding IP addresses belonging to major corporations, like Hewlett-Packard and government agencies, such as the U.S. Department of Defense. Once it identifies an insecure device, the malware tries to log in with a series of common default passwords used by manufacturers. If those passwords don't work, then Mirai uses brute force attacks to guess the password. Once a device is compromised, it connects to C&C infrastructure and can divert varying amounts of traffic toward a DDoS target.

Devices that have been infected are often still able to continue functioning normally, making it difficult to detect Mirai botnet activity from a specific device. For some internet of things (IoT) devices, such as digital video recorders, the factory password is hard coded in the device's firmware, and many devices cannot update their firmware over the internet.

The Mirai source code was later released to the public, allowing anyone to use the malware to compose botnets leveraging poorly protected IoT devices.

6.6 Malware

Malware is again software used to gain access to a device and infect them. Most of the IoT attacks are performed either by using a Trojan horse program or malware. BrickerBot attack and Mirai botnet are all created by adding a malicious software code to it. According to a report provided by Kaspersky lab [23], more than 8.5 million malware attacks have been performed during 2015 and 2016.

Why are these devices so vulnerable to malware infection? A number of reasons, but primarily because manufacturers have hastily created insecure products in their rush to benefit from the financial opportunities made abundant by inexpensive IoT technology. Under pressure to be competitive and quickly bring products to market, security has received very little attention. As a result, IoT devices commonly suffer from:

- *Weak authentication*: Passwords and login credentials are frequently left in their default state, many of which are weak and easily guessed. Some devices have solitary, fixed passwords, or virtually no authentication requirements whatsoever.
- *Numerous security vulnerabilities*: In many cases, products are designed by engineers with very little security expertise. History has repeatedly shown that all code has vulnerabilities. Software that's hastily developed or produced under extreme budget pressure has, even more, vulnerabilities.
- *Limited upgrade capabilities*: Inexpensive devices, like many IoT products, often have very low-profit margins, which can make it difficult or even impossible for manufacturers to afford to update firmware or send security patches.
- *Limited encryption*: A significant percentage of IoT devices are completely void of any encryption, either in transit or at rest.
- *Not on the security radar*: Not very many IT security personnel spend any energy regarding the security of smart thermostats, security cameras, DVRs, vending machines, or other "gadgets" connected to the company's network.

Malware infected smart gadgets are capable of inflicting harm in a number of ways, including the following:

- Denial of Service attacks

- Ransomware attacks
- Identity theft
- Account takeover
- Theft of IP.

It's time for enterprises to take IoT security seriously, and implement policies and tools to detect advanced malware that already has, or is attempting to establish a foothold in their organization. By investing a reasonable amount of time and effort to thwart IoT malware now, businesses will be much better prepared for the ever-increasing number of vulnerable devices that will surely be connecting to their networks.

7   IoT Security Solutions

In the previous section, we briefly discussed about the possible attacks that can be performed on an IoT device. In this section, we will discuss about the protocol stack of IoT architecture, various protocols that supports the architecture and the various solutions to enhance the security of IoT devices. The protocol stack of IoT [24] is as shown in Fig. 4.

Since, the Internet of Things consist of many connected devices such as sensors and RFID tags, it is important to adapt these devices to operate in a conventional internet. IoT devices are often constrained in computing power and memory capacity. Therefore it is a challenge to use cryptographic algorithms which often need more resources than the tiny devices have all together. Another challenge is updating devices in the field. There is often only an unreliable connection available and security critical things call for immediate updates, which can be difficult to roll out to all

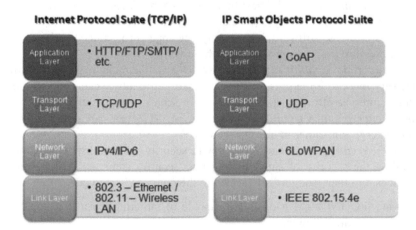

Fig. 4 IoT protocol stack

devices at once. Additionally the challenge of making security intuitive for the user is more relevant than ever, because the acceptance of users depends on easy installation and maintenance. Let me explain the protocols and security solutions that can be offered in IoT with various layers (Transport, Network and Application layers).

7.1 Transport Layer Solutions

The transport layer mainly involves two types of protocols. One is TCP and another one is UDP. In addition to these protocols, other protocols like Secure Socket Layer, Datagram Transport Layer Security and Quick UDP Internet Connections are explained to brief about the security in transport layer.

7.1.1 Transmission Control Protocol (TCP)

TCP is one of the widely used transport layer protocol where reliability is a major concern. TCP works on the principle of 3-way handshaking process. It has a connection establishment phase, data transmission phase and connection termination phase. This helps to achieve reliable data transfer. TCP is a connection oriented protocol. It determines how to break the application data into packets such that the network layer can easily process. Due to network congestion, some packets may get lost. TCP detects these problems and retransmits it.

The header format of the TCP [25] is as follows in the Fig. 5.

In this header, we have to separate field, namely checksum to ensure whether the received data is correct or not. But it doesn't provide any security mechanism to

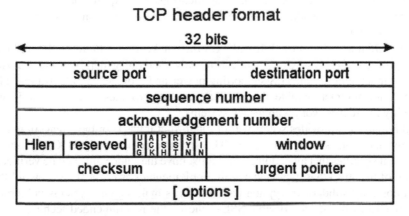

Fig. 5 TCP header format

prevent stealing of data. The security can be added with TCP in terms of SSL or TLS, which we will discuss in the subsequent sections.

7.1.2 Secure Socket Layer (SSL)

TCP does not provide any security to data. In order to transfer private data, SSL has been introduced. SSL uses a cryptographic system that uses two keys to encrypt the data—public key and private key.

When a Web browser tries to connect to a website using SSL, the browser will first request the web server identify itself. This prompts the web server to send the browser a copy of the SSL Certificate. The browser checks to see if the SSL Certificate is trusted—if the SSL Certificate is trusted, then the browser sends a message to the Web server. The server then responds to the browser with a digitally signed acknowledgement to start an SSL encrypted session. This allows encrypted data to be shared between the browser and the server.

Even though SSL provides security, it is still prone to Man-in-the middle attacks. To overcome the problems with SSL, we move to TLS.

7.1.3 Transport Layer Security (TLS)

SSL, or Secure Sockets Layer, is the predecessor to TLS, or Transport Layer Security. SSL has three versions, which are all considered insecure due to flaws in their design. TLS was created to address the weaknesses in the SSL protocol. The terms SSL, TLS and SSL/TLS are commonly used interchangeably in literature.

TLS is a protocol that provides privacy and data integrity between two communicating applications. It's the most widely deployed security protocol used today, and is used for Web browsers and other applications that require data to be securely exchanged over a network, such as file transfers, VPN connections, instant messaging and voice over IP.

Key differences between SSL and TLS that make TLS a more secure and efficient protocol are message authentication, key material generation and the supported cipher suites, with TLS supporting newer and more secure algorithms. TLS and SSL are not interoperable, though TLS currently provides some backward compatibility in order to work with legacy systems.

Although TLS provides security, it id found that it has kept the connection alive even when no data is being transmitted. TLS is not vulnerable to the POODLE attack, because it specifies that all padding bytes must have the same value and be verified, a variant of the attack has exploited certain implementations of the TLS protocol that don't correctly validate encryption padding. This makes some systems vulnerable to POODLE, even if they disable SSL—one of the recommended techniques for countering a POODLE attack. The IETF is working on the issue and still it is a draft.

7.1.4   User Datagram Protocol (UDP)

In contrast to TCP, yet another protocol namely UDP has been designed. It is a connectionless protocol. It has no handshaking dialogues, and thus exposes the user's program to any unreliability of the underlying network protocol. There is no guarantee of delivery, ordering, or duplicate protection. UDP provides checksums for data integrity, and port numbers for addressing different functions at the source and destination of the datagram.

When compared with TCP, UDP is preferred for IoT devices due to minimal overhead. In many resource-constrained embedded designs, UDP's lack of overhead makes a big difference in throughput when compared to TCP. UDP is connectionless and, therefore without a connection state to be maintained, so memory size/usage is not much of an issue. And because a UDP transaction requires only two UDP datagrams, one in each direction, load on the network is minimized, further reducing response times.

7.1.5   Datagram Transport Layer Security (DTLS)

DTLS is a communications protocol that provides security for datagram-based applications by allowing them to communicate in a way that is designed to prevent eavesdropping, tampering, or message forgery. The DTLS protocol is based on the stream-oriented Transport Layer Security (TLS) protocol and is intended to provide security guarantees. The DTLS protocol datagram preserves the semantics of the underlying transport—the application does not suffer from the delays associated with stream protocols, but has to deal with packet reordering, loss of datagram and data larger than the size of a datagram network packet

DTLS consists of two layers: the lower layer contains the Record protocol and the upper layer contains any of the three protocols namely Handshake, Alert, and Change Cipher Spec, or application data. The Change Cipher Spec is used during the handshake process to merely indicate that the Record protocol should protect the subsequent messages with the newly negotiated cipher suite and security keys. DTLS uses the Alert protocol to communicate the error messages between the DTLS peers. The Record protocol is a carrier for the upper layer protocols. The Record header contains among others content type and fragment fields. Based on the value in the content type, the fragment field contains the Handshake protocol, Alert protocol, change Cipher Spec protocol, or application data. The Record header is primarily responsible to cryptographically protect the upper layer protocols or application data once the handshake process is completed. The Record protocol's protection includes confidentiality, integrity protection and authenticity.

The DTLS Record is a rather simple protocol whereas the Handshake protocol is a complex chatty process and contains numerous message exchanges in an asynchronous fashion. The handshake messages, usually organized in flights, are used to negotiate security keys, cipher suites and compression methods. The scope of this paper is limited to the header compression only and not the cryptographic processing of Record and Handshake protocols.

7.1.6 Quick UDP Internet Connections (QUIC)

Quic is another multiplexed stream oriented protocol over UDP. Quic is designed to provide security equivalent to SSL/TLS. The main goal of this protocol is to improve the performance when compared with TCP.

The Key advantages of QUIC over TCP + TLS + HTTP2 include:

- Connection establishment latency
- Improved congestion control
- Multiplexing without head-of-line blocking
- Forward error correction
- Connection migration.

7.2 Application Layer Solutions

Internet is using HTTP protocol for a quite long time. Then what is the need for other protocols? HTTP is good for getting information by using request-response model. In the world of things, more devices are connected and there is a need for machine to machine communications. Also IoT devices keep on pushing information to the cloud or servers which it needs to send. In such a case, HTTP is not suited. Moreover, HTTP uses more bandwidth because of the text-based request and response model, which is not suited for low power bandwidth devices. Keeping in mind these things, two protocols has been developed one is MQTT and another is COAP.

Another reason for the popularity of these protocols is, they are smaller than HTTP, designed for machine to machine communications, Quality of Service and also tolerant to lossy networks.

7.2.1 CoAP (Constrained Application Protocol)

CoAP, Constrained Application Protocol, the name itself tells that is an application layer protocol. The application layer is just above the transport layer, where TCP and UDP are the basic protocols. The application layer protocols are built on any of these transport layer protocols (TCP or UDP). Basically TCP is complex when it is compared with UDP. The problem with UDP is that it is not stable. Since, HTTP is not suited for low power, low bandwidth devices; we are in a position to search for new protocols. At the same time, it is a successful standard for long time.

CoAP was proposed protocol in the year 2014. The developers of CoAP designed it in such a way that it should include the features of HTTP and also applicable for

Fig. 6 CoAP message structure

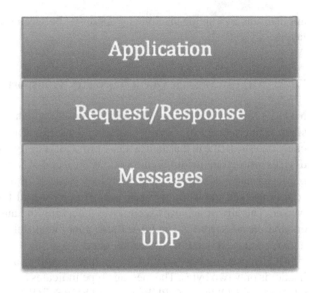

constrained devices. CoAP operates over UDP and is based on REST architecture. The CoAP message structure is shown in the Fig. 6.

CoAP employs a two layer structure, where the layers are Messages and Request/Response. The message layer comprises of CON (Confirmable), NON (non-confirmable), ACK (acknowledgement) and RST (reset). It is meant for retransmitting the lost packets.

1. CON (Confirmable)—when reliability is required, use this type of message. In this case, the messages are responded back with acknowledgement
2. NON (non-confirmable)—when reliability is not a big issue then use this
3. ACK (acknowledgement)—This type is to ensure reliability
4. RST (reset)—if something goes wrong, reset will be used.

The Request/Response layer contains methods like GET, PUT, POST and DELETE. CoAP protocol implements special features on HTTP which is not available in HTTP. The features are

(i) Observe flag—In HTTP, it is complicated to know the unused state on a variable. This flag is used along with GET message. Whenever there is a change in the observe flag, it will push the notification to the device
(ii) Discovery—This flag is related with discovering devices around us. The server can store the list of devices and the media types that they support.

The Quality of Service is achieved with the help of Confirmable and Non-Confirmable messages.

To protect CoAP transmissions, Datagram TLS (DTLS) has been proposed as the primary security protocol. Analogous to TLS protected HTTP (HTTPs), the DTLS-secured CoAP protocol is termed CoAPs. DTLS guarantees E2E security of different applications on a single machine by operating between the transport and application layers.

7.2.2 MQTT (Message Queue Telemetry Transport)

MQTT [26] is a TCP-based lightweight protocol which uses publish-subscribe messaging pattern (see Fig. 7). Any source such as a sensor can publish its data and any client can subscribe to that data. MQTT protocol is designed for resource-constrained devices whose bandwidth is minimal. MQTT consists of three components broker, publisher and a subscriber. The broker keeps track of all the publications and subscriptions. The publisher publishes information to all the subscribers through the broker. The broker achieves security by checking the authorization of the publishers and the subscribers. The broker also guarantees delivery of a message, i.e., it is delivered at least one time or exactly once.

MQTT uses binary format that requires a minimum of bandwidth. The fixed header is only two bytes. The message type indicates a variety of messages including CONNECT, CONNACK, PUBLISH and SUBSCRIBE. The DUP field indicates the message is duplicated and the receiver may have received it before. The QoS level field is for delivery assurance (see Fig. 8). MQTT supports three levels of QoS levels, "Fire and Forget", "delivered at least once" and "delivered exactly once". The RETAIN field informs the server to retain the last received Publish message.

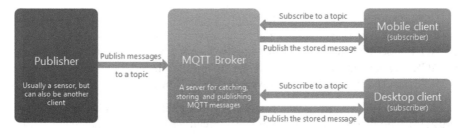

Fig. 7 Publish-subscribe model

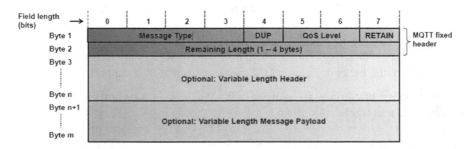

Fig. 8 MQTT message format

Not only it uses simple message format, but also requires less battery. MQTT was originally created in 1999 for remote sensors. It is now used for secure and reliable communication between devices. MQTT is based on Transmission Control Protocol (TCP) and can be secured with Transport Layer Security (TLS). MQTT provides minimal security. MQTT communications that rely on TCP alone are unencrypted and susceptible to man-in-the-middle attacks, DDoS attacks and buffer overflow attacks.

The next problem with MQTT is that MQTT messages are sent in clear text. Hence, the usernames and passwords are easy to access. To provide support to authentication process, it relies on Transport Level Security (TLS). Transport encryption with SSL and TLS can protect data when implemented correctly. To protect against threats, sensitive data including user IDs, passwords, and any other types of credentials should always be encrypted. The downside of using TLS, SSL, and other methods of encryption is that they can add significant overhead. However, techniques such as TLS session resumption can compensate for some of the connection costs of TLS. Hardware acceleration is another method for reducing the size penalty for encryption. For complex applications over constrained devices, an optimized encryption library can be very useful. When application code is large, an encryption library can reduce the processing memory and increase performance. The architecture of MQTT depends on brokers being highly available. Using X.509 certificates for client authentication can save resources on the broker side when many clients try to use broker services—such as database lookups or web service calls. Combining MQTT with state-of-the-art security standards like TLS and using X.509 certificates can also help improve security and performance. The encryption of data for safety and privacy is also critical to the revenue streams of service providers in delivering an optimal customer experience.

Even though MQTT is designed to be lightweight, it has two drawbacks for very constrained devices. Every MQTT client must support TCP and will typically hold a connection open to the broker at all times. For some environments where packet loss is high or computing resources are scarce, this is a problem. Also, MQTT topic names are often long strings. Both of these shortcomings are addressed by the MQTT-SN (MQTT—Sensor Networks) protocol, which defines a UDP mapping of MQTT and adds broker support for indexing topic names.

7.2.3 Secure MQTT (SMQTT)

MQTT and MQTT-SN both use SSL/TLS for security. But in reality, providing security certificates to all the devices is totally impossible. Also SSL/TLS suffers from attack such as BEAST, CRIME etc. To overcome this problem Secure MQTT which augments security feature for the existing MQTT protocol has been suggested by the author in [27]. Various message types are used in this protocol and are distinguished by message type in MQTT message header. Message type '0000' is reserved for future. Variable Header contains username and password flag (can facilitate user authentication), upon setting them, corresponding values are also included in payload. However, these values are not encrypted in the message and hence not secure. SMQTT protocol augments security feature to the existing MQTT by proposing a new MQTT Publish message Spublish with reserved message type '0000', where the messages encrypted using ABE (Attribute Based Encryption). Publisher uses Spublish command to publish an encrypted message using ABE. Hence, Subscribers who satisfy the access policy are capable of decrypting the message. The advantage of ABE is that it supports broadcast encryption which is suitable for IoT devices. ABE are of two types: (i) Ciphertext Policy based ABE (CP-ABE) and (ii) Key Policy based ABE (KPABE).

In secure MQTT protocol, there are three entities:

(i) Publisher device publishes the data under the given topic.
(ii) Subscriber device receives the data under the same topic through a Broker.
(iii) PKG or broker is the trusted third party.

There are four phases in the protocol. In setup phase, registration and key management are done. During encrypt phase, data is encrypted and in publish phase, Publisher publishes encrypted data under the given topic name and sends it to the broker. In decrypt phase, data is decrypted by subscribed devices.

7.3   Network Layer Solutions

The devices in the Internet of Things are resource constrained devices, which means the size of the device; the power and the memory capacity are limited. In this section, we will discuss about the IPv6 and 6LowPAN protocols. IPv6 is to support the address space of all the IoT devices involved. 6LoWPAN is specially designed for low power devices.

7.3.1   IPv6

IPv4 is the network addressing used widely. It is a 32-bit address and can support up to 4 billion devices. In the world of Internet of Things, every device is being connected to the Internet. But how is this possible? It can be done only when IP address is allocated to all the devices. IPv4 is not enough with such a huge number of devices. So, we move to IPv6.

IPv6 is a 128 bit address. It can allocate up to 2^{128} range of address. This makes it possible to allocate all number of devices that are connected in the IoT world. The major features that makes it advantageous over IPv4 is as follows:

1. Scalability—Since it is a 128 bit address, we can allocate IP address to every device
2. True end-to-end connectivity can be achieved
3. Address space utilization rates are small in IPv6

4. IP Sec is a requirement in IPv6, which allows two or more hosts to communicate in a secure manner by authenticating and encrypting each IP packet of a communication session.

7.3.2   6LoWPAN

Even though IPv6 provides the addressing platform, it is not suitable for the low power devices involved in IoT. To support these devices, we need another protocol. Low power Wireless Personal Area Networks (WPANs) which many IoT communications may rely on have some special characteristics different from former link layer technologies like limited packet size (e.g., maximum 127 bytes for IEEE 802.15.4), various address lengths, and low bandwidth. So, there was a need to make an adaptation layer that fits IPv6 packets to the IEEE 802.15.4 specifications. The IETF 6LoWPAN working group developed such a standard in 2007. 6LoWPAN is the specification of mapping services required by the IPv6 over Low power WPANs to maintain an IPv6 network. The standard provides header compression to reduce the transmission overhead, fragmentation to meet the IPv6 Maximum Transmission Unit (MTU) requirement, and forwarding to link-layer to support multi-hop delivery. Datagrams enveloped by 6LoWPAN are followed by a combination of some headers. These headers are of four types which are identified by two bits:

(00)—NO 6LoWPAN Header
(01)—Dispatch Header
(10)—Mesh Addressing
(11)—Fragmentation.

By NO 6LoWPAN Header, packets that do not accord to the 6LoWPAN specification will be discarded. Compression of IPv6 headers or multicasting is performed by specifying Dispatch header. Mesh addressing header identifies those IEEE 802.15.4 packets that have to be forwarded to the link layer. For datagrams whose lengths exceed a single IEEE 802.15.4 frame, Fragmentation header should be used. 6LoWPAN removes a lot of IPv6 overheads in such a way that a small IPv6 datagram can be sent over a single IEEE 802.15.4 hop in the best case. It can also compress IPv6 headers to two bytes.

References

1. Shipley AJ (2013) Security in the internet of things, lessons from the past for the connected future. Security Solutions, Wind River, White Paper
2. Jing Q, Vasilakos AV, Wan J, Lu J, Qiu D (2014) Security of the internet of things: perspectives and challenges. Wireless Netw 20(8):2481–2501
3. Roman R, Zhou J, Lopez J (2013) On the features and challenges of security and privacy in distributed internet of things. Comput Netw 57(10):2266–2279
4. Veracode white paper—The internet of things: security research study. https://www.veracode.com/sites/default/files/Resources/Whitepapers/internet-of-things-whitepaper.pdf/
5. Hajdarbegovic N (2017) Are we creating an insecure IoT? Secure challenges and concerns. https://www.toptal.com/it/are-we-creating-an-insecure-internet-of-things
6. Lewis N (2015) Prevent IoT security threats and attacks before its too late. http://internetofthingsagenda.techtarget.com/tip/Prevent-IoT-security-threats-and-attacks-before-its-too-late
7. Absolute security. https://www.absolutesecurity.co.uk/
8. Prevent enterprise IoT security challenges with preparation. http://internetofthingsagenda.techtarget.com/essentialguide/Prevent-enterprise-IoT-security-challenges-with-preparation
9. Johnson S (2017) Using mesh networking to interconnect IoT devices. http://internetofthingsagenda.techtarget.com/feature/Using-mesh-networking-to-interconnect-IoT-devices
10. Wheeler C (2017) Three new attack vectors that will be born out of IoT. https://www.liquidweb.com/blog/three-new-attack-vectors-will-born-iot/ [Three new attack vectors]
11. Higginis KJ (2017) IoT devices plagued by lesser known security hole. https://www.darkreading.com/cloud/iot-devices-plagued-by-lesser-known-security-hole-/d/d-id/1329320
12. Mah P (2008) New attack vectors challenge IT security pros. http://www.techrepublic.com/blog/it-security/new-attack-vectors-challenge-it-security-pros/
13. IoT standards and protocols. https://www.postscapes.com/internet-of-things-protocols/
14. Manoharan V (2016) TCP/IP layer-wise IoT protocols. http://www.synapt-iot.com/blog/tcpip-layer-wisc-iot-protocols/
15. LPWAN Technology Decisions: 17 Critical Features, Weightless STG (2016) http://www.weightless.org/membership/hvVs4ZGQqr5dwCDlBiYX
16. Protocol, W. Name: Azamuddin Rotation Project Title: Survey on IoT security. Chicago [Survey on IoT security]
17. Hossain MM, Fotouhi M, Hasan R (2015) Towards an analysis of security issues, challenges, and open problems in the internet of things. In: 2015 IEEE World Congress on Services (SERVICES). IEEE, pp 21–28
18. http://www.darkreading.com/iot/5-tips-for-protecting-firmware-from-attacks/d/d-id/1325604
19. http://safewayconsultoria.com/wp-content/uploads/2016/05/Beware-of-older-cyber-attacks_2016-1.pdf
20. https://www.welivesecurity.com/2016/10/24/10-things-know-october-21-iot-ddos-attacks/
21. https://www.enisa.europa.eu/publications/info-notes/major-ddos-attacks-involving-iot-devices
22. http://krebsonsecurity.com/
23. www.symtrex.com/category/iot
24. Sutaria R, Govindachari R (2013) Making sense of interoperability: protocols and standardization initiatives in IOT. In: 2nd International Workshop on Computing and Networking for Internet of Things
25. NetworkWolves (2015) https://networkwolves.wordpress.com/2015/03/20/tcp-and-udp-and-difference-between-them/
26. MQTT basics in IoT. http://www.rfwireless-world.com/Terminology/MQTT-protocol.html
27. Singh M, Rajan MA, Shivraj VL, Balamuralidhar P (2015) Secure mqtt for internet of things (iot). In: 2015 fifth international conference on Communication Systems and Network Technologies (CSNT). IEEE, pp 746–751
28. Connected Consumer Product Best Practice Guidelines (2016) IoT security Foundation. https://iotsecurityfoundation.org/wp-content/uploads/2016/12/Connected-Consumer-Products.pdf

Security of Big Data in Internet of Things

Rakesh Bandarupalli and H. Parveen Sultana

Abstract Presently 25 billions of devices are connected to the internet, and 50 billions of devices will connect to the Internet by 2020. These devices comprise of lots of sensors. These sensors, computers, tablets, and smart phones are generating twice as much as the data today as they generated two years ago. According to a survey, 90% of connected devices are collecting the information, and 70% of this data is transmitting without encryption. The rapid growth of the data brings tremendous changes in the humans' daily life. This data is providing new business opportunities. Many IoT devices are generating data related to the personal behavior. Lots of research is happening in big data security. This research is in the initial stage only. Up to right now, there is no specific method for providing security to the big data. Most of the data generated with IoT applications is unstructured data. Providing security to the unstructured data is more difficult than the providing security to the structured data. This chapter discusses various mechanisms for providing security to the big data generated by various IoT devices. This chapter describes the existing techniques for providing security at data generation, transmission and storage phases. The first methodology describes security to the data on Internet of Vehicles. This methodology uses a single sign-on algorithm. In this technique vehicle node and slink nodes need to register at a big data center for one time. This technique uses symmetric key cryptographic algorithms for encrypting the data. The second methodology describes providing the security with a dynamic prime number based security verification scheme. In this methodology, the prime numbers will be generated at regular intervals of time. The prime numbers will be generated at both source and receiver side. 128-bit symmetric key cryptography is used for this methodology. This paper also discusses about the advantages and disadvantages of these methodologies.

R. Bandarupalli
Sree Vidyanikethan Engineering College, Tirupathi, Andhra Pradesh, India
e-mail: bandarupallirakesh@gmail.com

H. Parveen Sultana (✉)
VIT University, Vellore, Tamil Nadu, India
e-mail: hparveensultana@vit.ac.in

© Springer Nature Switzerland AG 2019
N. Jeyanthi et al. (eds.), *Ubiquitous Computing and Computing Security of IoT*,
Studies in Big Data 47, https://doi.org/10.1007/978-3-030-01566-4_2

Keywords Internet of Vehicle (IoV) · Secure mechanism · Big data · Large-scale IoT · Big data analytics

1 Introduction

The computational capability is increasing drastically from the past decade. The development of networking is leading towards the rapid growth of the web technologies and data centers [1]. The development of Internet of things and big data is quickly accelerating and impacting all areas of technologies. This enhancement is benefiting many organizations as well as individuals [2]. The IoT devices are playing a vital role in this enhancement. The Big Data can be categorized depending upon three factors Velocity, Volume, and Variety. Gartner introduced these terms to describe the challenges of big data [3]. Massive opportunities are producing by analyzing this IoT data in the domain of smart cities, smart transportation, health care and much more.

The rapid growth of IoT devices triggers big data analytics as a challenging task. According to the estimation of IDC (International Data Corporation), the big data market will reach more than the US$125 billion by 2019. The big data analytics used to extract the useful information using various data mining techniques [4]. This information is useful in taking the business decisions as well as in revealing the recent trends.

The IoT data is different from the big data collected through the systems. As the IoT data is collected from various sensors, this data consists of a lot of noise, heterogeneity, and variety [1]. Various studies said that the sensors will increase to 1 trillion by 2030. This enhancement will cause the producing of huge amounts of data [3]. The area like smart traffic, smart grids, intelligent logistics management and intelligent buildings are the some of the applications of IoT and Big Data.

Big data is a term refers to large data sets. These data sets are complex in nature. The traditional data processing applications are not suitable for processing the data [2]. The big data consists of both structured as well as unstructured data. The data is generated from various sources like social networking sites, health care applications, and sensor networks and from many other organizations.

Internet of things is envisioned as the emerging trend. There is a lot of scope for research. This technology makes the human life comfortable. It shows solutions to the lot of problems related to logistics, transportation, urbanization, and environment. This technology enables to connect the physical world things and cyber world.

2 Introduction to Internet of Things and Big Data

2.1 *Internet of Things*

IoT is a platform where the devices are communicated with one another over the internet. These devices consist of various types of sensors. The IoT devices will share the information in a convenient manner. The IoT is termed as the next generation

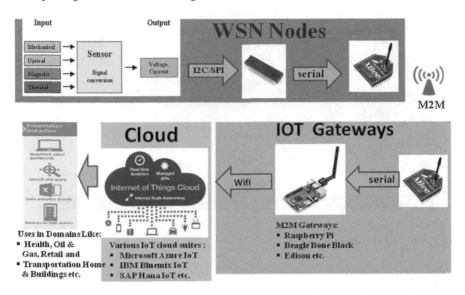

Fig. 1 Internet of Things big picture

revolution. IoT is adopted in various sectors such as smart cities, smart transportation, smart office, smart retail, smart energy and smart health care.

Figure 1 is representing the Internet of Things Big picture. It consists of various communication mechanisms used in IoT, Different IoT gateways, different storage mechanisms used for storing the data produced by IoT devices and also various applications that use Internet of Things.

The mobile devices, transportation vehicles, home appliances, health care devices etc. are used for data actuation [3]. Wrist watches, Doors, refrigerators, air conditioners and microwave Owens are the some of the IoT devices. These devices are deployed in various geographical locations [5]. These devices will acquire the real time data. These devices are connected with several communication mechanisms such as Bluetooth, Zigbee, WiFi etc. 50 billion devices such as laptops, smart phones, sensors will connect to the internet.

The graph in Fig. 2 is generated from the data provided by Cisco in the year of 2011. That graph is representing how the IoT devices are increasing rigorously in this decade.

Figure 3 is representing that by 2020 50 billion connected devices will be available for 7.6 billion people, and also the above figure is showing that the IoT is the combination of people, devices, and sensors interconnected with one another.

2.2 Big Data

The IoT devices and various other software applications will produce the data continuously. This data will consist of Structured, Unstructured and Semi structured data.

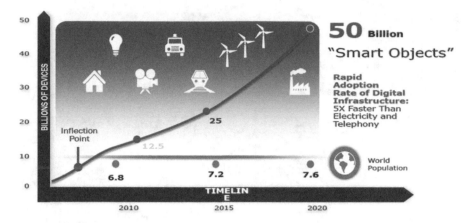

Fig. 2 Growing of Internet of Things

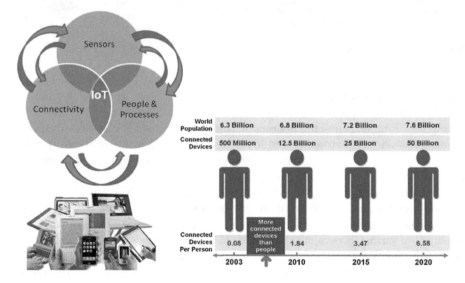

Fig. 3 Connected world

This huge amount of data is termed as "Big Data". The conventional data bases are not sufficient to store this huge amount of data [6]. In simple terms, the big data can be defined as the data that cannot be handled by the single system. The conventional data bases cannot be used for processing and analyzing this data that is growing rigorously.

Gartner proposed a model consisting of 3V's (Volume, Velocity, Variety). In other words Volume of the data producing, Variety of the data producing, Velocity or speed of the data producing [1]. Some investigations said that the volume is the main characteristic of the big data.

2.3 Big Data Analytics

The big data analytics examines large data sets consisting of a variety of data to provide useful business information, market trends. By analyzing this huge amount of data can help the organizations in getting the useful information. Big data analytics require tools and technologies that can transform structured, unstructured and semi structured data into more useful data [4–6]. The scientists can analyze large volumes of big data using the traditional tools.

2.4 Relationship Between Big Data Analytics and IoT

The big data analytics are used for decision making by analyzing the data produced by IoT devices continuously. The big data analytics are used to analyze the continuous data and store this large amount of data using various storage technologies. This large amount of data mostly consists of unstructured data [7]. Here the analytic tools need to analyze this data with lightning speed so that the business organizations can take the decisions immediately. Need of adopting big data in Internet of Things applications are increasing dramatically. Figure 2 is representing how the big data and IoT are interdependent on one another. As the usage of IoT devices is increasing the use of the big data will also increase proportionally [8–10]. The combinations of these two technologies are providing good business opportunities in the area of business and research.

Figure 4 is representing that how the big data analytics and Internet of Things are inter connected with one another. The Figure consisting of three phases, the first phase consists of IoT devices with sensors, these devices are interconnected with one another [11–14]. The second phase consists of different storage technologies. The data produced by IoT devices are stored on low-cost commodity hardware. This data can be called as big data, this data has mainly three properties i.e. volume, velocity, and variety. This data will be distributed among fault tolerant databases. The third phase is an analytical phase. In this phase, various tools will be used for analyzing the data such as MapReduce, Spark, Skytree, and Splunk. These tools require training data set. With the help of training data sets we use queries, then produce reports and result sets.

2.5 Architecture of IoT for Big Data Analytics

The architecture in Fig. 5 is representing the Architecture of Internet of thing for Big data analytics. This architecture is consisting of seven layers. The first layer consists of IoT devices which are having sensors. The second layer consists of communication devices such as Internet, Zigbee, WiFi, and Bluetooth etc.

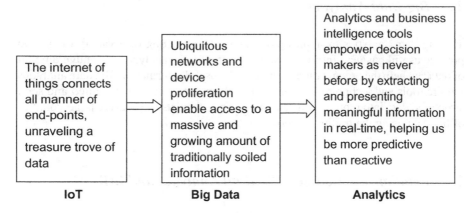

IoT	Big Data	Analytics
The internet of things connects all manner of end-points, unraveling a treasure trove of data	Ubiquitous networks and device proliferation enable access to a massive and growing amount of traditionally soiled information	Analytics and business intelligence tools empower decision makers as never before by extracting and presenting meaningful information in real-time, helping us be more predictive than reactive

Fig. 4 Relationship between Internet of Things and big data analytic

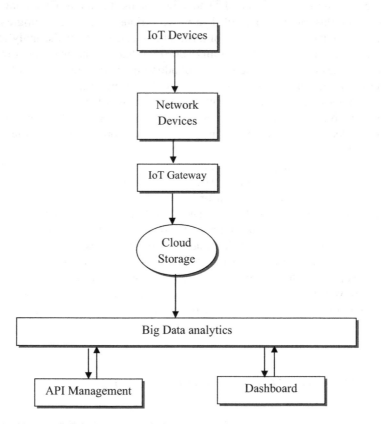

Fig. 5 Architecture IoT for big data analytics

The next layer is the cloud which is constructed with commodity hardware. The data generated by IoT devices will be stored in this cloud. The data will be received to the cloud through IoT gateway. The next phase consists of big data analytics phase. In this phase, a large amount of data will be processed which is stored in the commodity hardware. The major purpose of this architecture is to provide ample business solution.

3 Privacy of Big IoT Data

Sensitive information (Personal details) of users will be grabbed by the IoT devices. So, security is one of the major issues in the Big IoT data. These systems majorly depend on third party services. The traditional security solutions are not much effective in protecting this huge amount of data. The current existing security algorithms are majorly designed for providing security to the static data. The data produced from IoT devices are dynamic data. The data can be protected at generation phase, data storage phase, and data processing phase. Information privacy is protecting the information of a person or an individual from others. Security is protecting the data by using technology from recording, modifying, deleting.

3.1 Big Data Privacy at Data Generation Phase

The big data can be generated passively or actively. In active data generation, the data generated by the user will give to the third party. In passive data generation, the data will be generated by the user and the user will not have awareness about either the data is collected by the third party or not. The data can be protected at data generation phase either by access restriction of data or by falsifying the data.

a. **Access Restriction**

In most of the cases user not interested to share the sensitive information. If the user wants to share the data passively the user will take some precautions to secure the data by blocking the advertisements, blocking the scripts, and also by using some encryption techniques.

b. **Falsifying the Data**

In several cases it is very difficult to protect the sensitive information; in such cases, data falsification is used. In data falsification, the data will be distorted by using various tools. For example, while us using the credit card for online shopping Mask Me tool is used by most of the merchants.

3.2 Big Data Privacy at Data Storage Phase

The enhancement of big data technologies is leading to overcoming the storage problem. But if the big data storage system is compromised in security aspect, it will lead to a disclosure of the Users personal information. There are four categories in the traditional security mechanism. They are data security schemes at the file level, data base level, medium level and encryption scheme at the application level. The big data infrastructure should be scalable. By using storage virtualization we can accommodate more than one application dynamically. In this storage virtualization, more than one network storage devices are combined dynamically, so that we can assume that this is a single storage device. The data storage security and also computation auditing security can be provided with the help of SecCloud model.

3.3 Privacy Preservation Approaches for Cloud Storage

There are mainly three factors to be considered in storing the data securely n the cloud, i.e. integrity, confidentiality, and availability. The integrity and confidentiality are directly related to the security aspect of the data. The availability is representing the authorized persons can access the data whenever they required it. There are some basic methods to fulfill the security aspect of the data [15]. For example, the sender will encrypt the data with a public key and the receiver will decrypt the same data using a private key. The mechanisms for ensuring the privacy of the data are Attribute based encryption, Storage Path encryption, Homomorphic encryption and Using of Hybrid clouds.

3.4 Verification of Integrity of Data in Big Data Storage

When the data is stored in a third party cloud, the user will not have control over the data. So the data is at risk. In this scenario, the user needs to verify whether the data is stored in the cloud or not properly [16, 17]. This verification is called for checking the integrity of the data. To verify the integrity of the data there are several mechanisms provided. They are Message authentication code, Digital signatures, Checksums, trap-door hash functions, and Reed-Solomon code. We can also verify the integrity of the data available in the cloud by retrieving all the data stored in the cloud. The integrity verification is having the highest priority in security aspect.

3.5   Privacy Preserving of Big Data in Data Processing

Batch processing, machine learning, stream processing and graph processing are the big data processing paradigms [18, 19]. We can provide security to the data in two phases. In the first phase, the data should be protected from disclosing to the others. If the data is disclosed then the personal or sensitive information of the user will be at risk. In the second phase, the meaningful information needs to be extracted from the data without violating the privacy.

4   A Secure Mechanism for Big Data Collection on Internet of Vehicles

Internet of Vehicles is an extension of Internet of Things. The internet of vehicles is under smart transportation domain. On Internet of Vehicles, the vehicles get to connect with one another and also with the internet. This connection is leading producing of the data of different dimensionalities [20]. This data consists of the vehicles' location, a speed of the vehicle and the route in which the vehicle was traveled. This type of information will be collected by different sensors of the vehicles. The analysis of this information carries huge research interest. This research will be useful in traffic management [21]. This information may also consist of the users' personal information. If the security is not provided for this data then users' privacy will be at risk. There may be a chance that fraudulent information may be transmitted by the malicious vehicles to disturb the traffic system intentionally. So it is necessary to take the precautions to avoid the malicious vehicles.

Here in this security mechanism, first of all, the vehicles need to register at the big data centers to authenticate the vehicles. This authentication will be done by using the single sign-on algorithm. The basic architecture of the internet of vehicles is as represented in Fig. 6. This architecture consists of 4 blocks, i.e. Vehicle nodes, Road side units, Big Data centers and Storage module. This architecture also includes Satellite [22]. The vehicle nodes will communicate with one another and also with the Road side units. These road side units are also called as the SLINK nodes. These vehicles will be communicated with the help of the internet also, for that purpose the vehicles also interact with the satellites. The data generated with the help of vehicles will be collected by the SLINK nodes. This data will be transferred to the big data centers, where the collected data will be analyzed and again the information will be transmitted to the vehicle nodes. This information consists of in which route the traffic is huge and in which route the traffic is less [23]. The data will be finally stored in the storage module.

The diagrammatical representation of this scheme is shown in Fig. 7. As the no of vehicles increases, the data of different attributes will also increase. This data will be collected from different geographical locations. This data will be stored in the big data centers. These centers are distributed storage systems. These centers use Hadoop

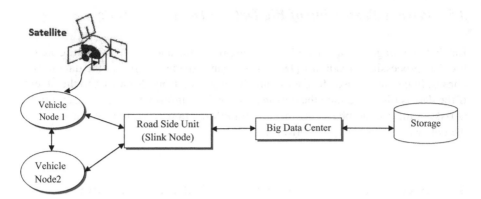

Fig. 6 Architecture of internet of vehicles

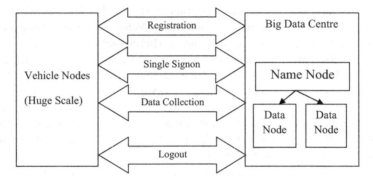

Fig. 7 Collecting data in a secure manner

Architectures. In the first phase authentication of all vehicle nodes will be done. Here the vehicle nodes will register at big data centers and required information will be exchanged with the big data centers. After registration phase, the vehicles will be log-on to the big data centers using a single sign-on algorithm. After that, the data will be exchanged continuously until up to the vehicles logouts from the system.

In this methodology, all the vehicles will register at the big data center for entering into the network. In the second phase authentication will be done by using a single sign-on algorithm [24]. In the next phase the collected information will be transferred securely and efficiently. In the final phase the collected information will be stored using distributed storage.

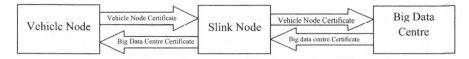

Fig. 8 Exchanging of messages at initialization phase

4.1 Initialization Phase

Here each and every vehicle is equipped with a certificate given by a third certification authority. In this phase, the vehicles need to register with the big data centers. The vehicles and big data centers generate a public key and private key among themselves. As shown in the Fig. 8 certificates and public keys will be exchanged in between the Vehicle node and big data center. The slink node acts as a mediator [25]. After that the certificates will be verified then the vehicle will get registered in the big data centers.

4.2 First Time Log-on

This section describes different procedures for slink node and vehicle login using single sign-on algorithm.

In the slink nodes' sign-on phase ID of the slink node, random number to fight against the replay attack, Message time stamp and slink nodes signature will be sent to the Big Data center. The big data center will check all these details. If these messages are from valid slink nodes then the big data centers will generate a session key. This session key is a unique key. The big data center will forward a packet consisting of a random number and unique session key (sc_key). This packet will be encrypted with the public key of the slink node. The slink node will decrypt the private key and acquires session key. Table 1 list out the symbols used in Fig. 9.

Table 1 Symbols notations

Symbol	Description
Rno	The random number to fight against replay attack
Tstamp	Message time stamp
Slink_Sign	Slink nodes signature
slink_Pk	Slink nodes public key
sc_key	Session key between big data center and slink node

Fig. 9 Logging on of slink nodes' for the first time

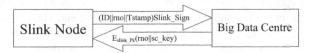

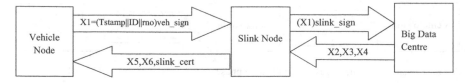

Fig. 10 Vehicle nodes' first-time log-on

In the Vehicle nodes' sign-on phase, the vehicle node sends the ticket to the slink node with its signature. The ticket has a time stamp of the message, vehicle node ID and random number generated by the slink node for fighting against the replay attack. Then the slink node sends the same ticket to the big data center [26]. This ticket consists of signature of the slink node. Big data center validates these details and three tickets will be passed to the slink node. The first ticket consists of Time stamp, ID of the big data center and a Random number generated by the big data center along with the big data center signature (See Fig. 10).

$$X2:(Tstamp\|ID\|rno)cen\_sign$$
$$X3:E_{veh\_pk}(vc\_key)$$
$$X4:E_{sc\_key}(X2\|X3)$$
$$X5:E_{vs\_key}(vs\_key)$$
$$X6:E_{veh\_pk}(X2\|X3)$$

The second ticket consists of session key between vehicle node and big data center. This session key is encrypted with vehicles public key [27]. The third ticket consists of both first and second tickets these tickets are encrypted with session key between slink node and big data center. The slink node generates a session key between vehicle node and slink node. The X2 and X3 are encrypted with this session key and this packet will be forwarded to the vehicle node. The session key also forwarded to the vehicle node by encrypting it with the public key of the vehicle. Table 2 has the description of various symbols used in this scheme.

4.3 Once Again Log-on

As the vehicle nodes are in the moving condition, the vehicle nodes need to log-on to the next arriving slink nodes by leaving the current log-on slink node. When the vehicle nodes want to access another slink node by leaving the first log-on slink node we need to follow the scenario discussed in this section [28]. Figure 11 is representing the communication between the slink node and vehicle node. In this communication process session key will be updated. The stored ticket X2, vehicle certificate will be forwarded to the Slink node with vehicle signature.

Table 2 Symbols notations of vehicle node

Symbol	Description
Tstamp	Time stamp
Rno	Random number for fighting against replay attack
Veh_sign	Vehicle nodes signature
Slink_sign	Slink nodes signature
Cen_sign	Big data centers signature
Veh_pk	Vehicle nodes public key
Vc_key	Session key between big data center and vehicle node
Sc_key	Session key between big data center and slink node
Vs_key	Session key between slink node and vehicle node
Veh_pk	Vehicle nodes public key

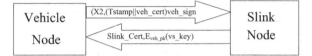

Fig. 11 Later log-on of vehicle node

The ticket X2 consists of the big data center signature. This signature shows that the ticket is issued by the big data center. After that, the session key (vs_key) of the slink node will be encrypted with the vehicle nodes public key. This session key (vs_key) will be forwarded along with the slink node certificate (Slink_cert).

4.4 Collecting Data Securely

In the previous scenarios the secure connection will be established between the Vehicle node, Slink node and big data centers. The data will be divided into two categories, business data, and confidential data. The business data will be exchanged in the plain text format and confidential data will be exchanged securely [29]. The business data consists of the information like temperature. The X4 can be calculated by concatenating vehicle nodes' Id with business message M1. The hash value of M4 is utilized for calculating HMAC. HMAC helps in stop the tampering of data. So that, the data will be sent to the receiver without any loss. The same scheme will be used to transfer the data from a big data center to the Vehicle node. Here M2 is the business message to be sent from the big data center to the vehicle node (See Fig. 12).

But the confidential data need to be exchanged securely. So here we are encrypting the confidential data and converting into the cipher text format. Here a random key, Z is utilized for encrypting. For sharing the random key Z with the slink node and big data center, vc_key and vs_key will be used (See Fig. 13).

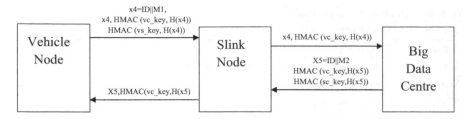

Fig. 12 Exchanging of business messages for big data collection

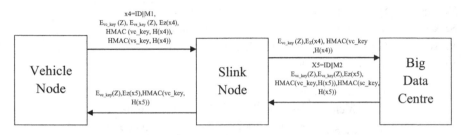

Fig. 13 Exchanging of Confidential messages for big data collection

4.5 Data Storage Security

In the previous sections we have discussed the secure connection establishment and secure data collection [30]. This section describes the storing of important data in big data center securely. All the information related to the vehicle need not be stored in the big data center. Some important information needs to be stored in the big data center securely [31]. Table 3 is the data structure of the information need to be stored in the big data center. It has the data structures related to both slink node and vehicle node. The first field of the data structure is ID, the second field is the certificate. This ID and certificate fields are used for identification purpose. The third field is statue field. This field consists of two values "on" and "off". If any abnormal situation occurs, the status field will change from on to off. The next field is the time stamp period field. If the time stamp period expires the vehicle node and slink nodes need to register once again as the new nodes. The session key and public key are important in providing confidentiality to the data.

The business information will be stored as a plain text in the big data center [32]. While storing the confidential information, the data need to be encrypted with the

Table 3 Data structure of big data center for storing slink node and vehicle node data

Nodes	ID	Certificate	Statue	Validity period	Encrypted session key
Slink node	Slink_ID	Slink_cert	Off/on	Period_TStamp	Sc_key
Vehicle node	Veh_ID	Veh_cert	Off/on	Period_Tstamp	Vc_key

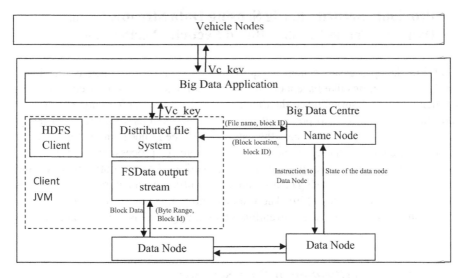

Fig. 14 Distributed storage system for big data collection

session key between vehicle node and big data center (vc_key). When the vehicle node itself interacts with the big data center, the data will be retrieved [33]. Otherwise, the data cannot be retrieved from the big data center. When the vehicle node interacts with the big data center the data will be decrypted with the session key between vehicle node and big data center (See Fig. 14).

As the no of vehicles is increasing day to day, the data collected from the vehicles also increasing rapidly. The Hadoop Distributed File System (HDFS) is a famous system for storing the big data. In this HDFS system we will have one name node and remaining all are data nodes. In the HDFS system the data will be replicated to more than one location to avoid the fault tolerance. Whenever a vehicle node wants to access the data, the vehicle node interacts with the big data center, the client JVM request for the file name and block ID through the distributed file system [34]. This distributed file system interacts with the Name node. The name node will acknowledge the block location and block id to the distributed file system [35]. Finally, FS Data output stream sends the Block ID and byte range to the data node to acquire the data [36]. If the acquired data is business data then the data will be sent to the vehicle node as a plain text. Otherwise, the data will be sent as the cipher text. The cipher text needs to be decrypted by using the session key between the vehicle node and big data center (vc_key).

5 Providing Security to Big Sensing Data Streams Using Dynamic Prime Number Based Security Verification

Real time data processing schemes require in many applications such as social networking applications like Facebook and Twitter, large scale sensors, web exploring, financial data and surveillance data analysis [37]. Stream processing engines are introduced with an aim to process the sensing data streams with the small delay. These engines are used to process the data in real time rather than processing after storing the data. But these engines are not suitable for processing the big stream data [38]. These large quantities of data contain various data, i.e. both structured and unstructured data. As these big data streams are continuous in nature, this data needs to be processed in real time [39]. The velocity and volume of this data are huge, we cannot store this data. So, the conventional computing models are not suitable.

5.1 Security Verification of Data Streams

The big sensing data streams are used in some critical applications such as military, these data need to be secured. The sensors are having the low processing power, less power, low storage and also very less energy [40]. These data streams need to be processed during the transmission phase itself. Here providing the security to the data is a very important aspect. For providing security to the data cryptographic model is used. There are two cryptographic models such as Asymmetric and Symmetric [41]. The asymmetric algorithms are much slower when compared to the symmetric cryptographic algorithms. But these symmetric cryptographic algorithms are failed in many cases when providing the security to the streamed data.

As the symmetric key cryptographic algorithms are failed in many cases of big data streaming, the dynamic prime number based security verification scheme will address those challenges. In this scheme, the key will be generated with the help of prime numbers synchronously. This key is generated at regular intervals of the time [42]. This prime number generation will be done at both sensing device side as well as at data stream manager side. As the key is generated at both source and DSM sides, it reduces communication overhead [43]. Here the key is of 64-bit size. This smaller key helps in faster processing of the streamed data by not compromising the security. The key is updated dynamically at both source and DSM side.

5.2 Architecture of Secure Data Stream

a. **Data Stream Processing**

The data stream processing is a revolutionary area. Many applications are using this data stream processing. In data stream processing huge amounts of data need to

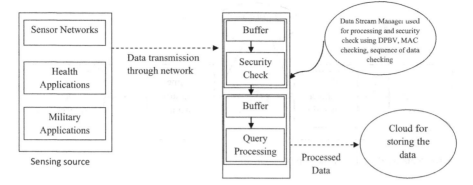

Fig. 15 Architecture of data stream

analyze with a small delay. In conventional mechanisms data is analyzed after storing it [44]. Here the data will be generated from various sources. It is very difficult to handle the data generated from various sources at a time [45]. And also in DSM, the data blocks need to go under the security verification.

The Fig. 15 is representing the architecture of secure data stream. In this architecture, the data stream flows from various sensor devices to the cloud. Here the architecture mainly focuses on three aspects, collecting the data, processing the data and storing the data. The security and query related processes are done in DSM (data stream manager). In this architecture first of all security verification will be done after that query processing will be done [46]. Small buffers will be maintained for both activities. In the final stage the data that is processed will be stored in the cloud [47]. Here the queries used for processing the data are continuous in nature, as the data is flowing continuously.

5.3   Purpose of Symmetric Key Cryptography

The size of symmetric keys is much smaller in size when compared with the asymmetric keys, so they require less computation power. A 128-bit symmetric key provides the equal strength compared with the 3240-bit asymmetric key. The main aim of big sensing data streams is to provide security to the data streams in real time. So, the symmetric key cryptography is the best choice in this scenario. The symmetric key cryptography is 1000 times faster than other public key cryptographic algorithms. As the size of the symmetric key cryptography is much smaller, the attacker can easily attack the data which is encrypted with symmetric key cryptography. To overcome this disadvantage, the keys are generated synchronously with dynamic number based algorithm [48]. Here the keys will be generated at both sensing devices end as well as at Data Stream Manager at regular intervals of time. The Fig. 16 will demonstrate about this key generation scheme.

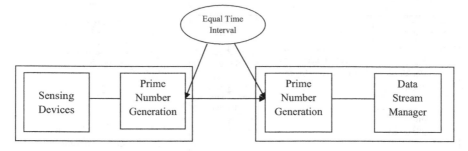

Fig. 16 Relative dynamic prime number generation

5.4 Setup of DPBSV System

Here the system is completely untrusted. The Data stream Manager (DSM) should maintain the entire sensor ID's and also secret keys (See Table 4 and Fig. 17).

Here in this process, first of all, sensors ID and a pseudo random number will be sent to the DSM. The DSM receives these details from Sensor. After that, the DSM retrieves secret key (key_s), with the help of Retrieve key function. After that, the session key (key_{si}) will be generated with the help of random key function. This session key will be combined with the secret key. This combination will generate a key (Key_{enc}) used for authentication purpose. The generated key and session key will be encrypted with the shared key (key). The hash value 1(H) will be computed with the help of hash function. The computed hash value and sensors private key will be passed to the sensor. The below given are the steps for computing hash value1.

Table 4 Symbols notations and descriptions

Symbol	Description
SID	Sensor's ID
Rno	Pseudo random number
Key_{enc}	Key generated for authentication
$H/H'/H''$	Computed hash value
Enc()	Function used for encryption
Key	Shared key initially used for authentication of DSM and sensor
Key_d	DSM secret key
It	Prime number generation interval time
RP	Prime number generated randomly
PF(RP)	Function used for generating the prime number randomly
Key_{sh}	DSM and sensor computed secret key
Key_s	Secret key of the sensor
M/M'	User authentication key which is encrypted with sensors' secret key

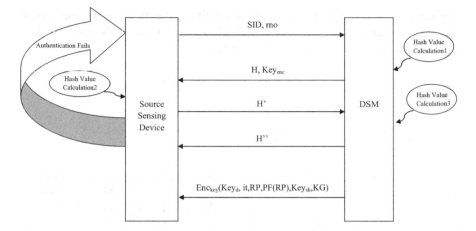

Fig. 17 Secure authentication procedure between DSM and source sensing device

$$\text{Key}_s \;\text{<-}\; \text{Retrieve (SID)},$$
$$\text{Key}_{si} \;\text{<-}\; \text{random ()},$$
$$\text{Key}_{enc} \;\text{<-}\; \text{Key}_s \oplus \text{Key}_{si} \tag{1}$$
$$M = \text{Enc}_{key}\left(\text{Key}_{si}, \text{Key}_{enc}\right)$$
$$H = \text{Hash}\left(\text{Key}_{enc}\|M\|\text{rno}\right)$$

After that, the sensor will get the hash value (H) and key used for authentication purpose (key_{enc}). The DSM will get these details and finds its own secret key based on the authentication key. The sensors secret key and authentication key (Key_{enc}) will be encrypted for users' authentication. The hash value will get with the equation Hash ($\text{Key}_{enc}\|M'\|\text{rno}$). And validates whether the hash value generated by it and the hash value generated by DSM are equal or not. If $M = M'$ and the hash values are equal, then the authentication of DSM is successful by the sensor. If the authentication process is failed then again the process will begin from step 1.

$$\text{key}_{si} = \text{Key}_{enc} \oplus \text{Key}_s$$
$$M' = \text{Enc}_{key}\left(\text{key}_{si}, \text{key}_{enc}\right)$$
$$\text{Hash}\left(\text{key}_{enc}\|M'\|\text{rno}\right) \tag{2}$$
$$M = M', \text{ for authentication of DSM}$$
$$H' = \text{Hash}\left(1\|\text{Key}_{enc}\|M'\|\text{rno}\right)$$

After that the H′ will be forwarded to the DSM, the DSM compares the received value with the Hash(1‖Key_{enc}‖M‖rno), If these two are equal then the sensor is authenticated successfully. If the authentication is failed then the protocol will be terminated. In this way, both the sensor and DSM authenticate each other. After successful validation, the DSM sends another hash value to fulfill the protocol. The hash value H'' will be calculated as shown below.

$$H'' = \text{Hash}\left(2\|\text{Key}_{enc}\|M\|rno\right)$$

5.5 Handshaking of DPBSV

During the calculation of prime numbers, we need to take care of communication overhead. The communication overhead must be reduced. The PF(RP) function used to generate the prime numbers randomly at both sides. These prime numbers have to be generated at regular intervals of the time. The DSM transmits the algorithms related to generating of the prime number and keys like (Key$_d$,it,RP,PF(RP),KeyGen, Key$_{sh}$) to each and every individual sensor by encrypting them with the shared key generated initially. This transferred information will be stored in the trusted part of the sensor.

After successful completion of handshaking process, the data needs to be transmitted securely. This secure transmission and verification can be done by using several functions and keys. As discussed earlier this scheme utilizes dynamic prime number generation process. This dynamic prime number generation can be done at both sensor and DSM side. Each and every sensor will have its own key. Initially shared key and prime numbers will be generated by the DSM itself. Next prime number will be generated depending upon the current prime number and the interval time. The shared key will be generated by the sensors depending upon the formula $\text{Key}_{sh} = \text{Hash}(\text{Enc}(\text{RP,key}_d))$. Here, each and every data block consists of two parts. The first part consists of the encrypted data. This data will be encrypted with the help of secret key Key$_i$ and shared key key$_{sh}$. These three things will be mutually exclusively ORED. DATA $\oplus$ Key$_i$ $\oplus$ key$_{sh}$. This encryption is mainly used for integrity checking. The second part is used for authentication checking. $S_i \oplus$ key$_{sh}$. So finally the resultant block is

$$(\text{DATA} \oplus \text{Key}_i \oplus \text{key}_{sh})\|(S_i \oplus \text{key}_{sh.})$$
$$\text{Lets } I_d = \text{DATA} \oplus \text{Key}_i \oplus \text{key}_{sh}$$
$$A_d = S_i \oplus \text{key}_{sh}$$

In the next step the sensor will send the encrypted format of the above data block

$$\text{Enc}_k(A_d\|I_d)$$

5.6   Security Verification of DPBSV

The security verification should be done in real time. The main aim of the security verification is to provide the end to end security. This security verification will be done at DSM side. In the security verification the DSM verifies whether the data is modified or not. And also it verifies whether the data is from the authenticated node or not. First of all the DSM will decrypt the data block to check the integrity and authenticity. First of all the DSM authenticates each and every block. And the integrity will be checked at the arbitrary interval blocks. The interval may vary from 0 to 6. i.e. the interval blocks may be 6 at most or 0 at least.

The Fig. 18 representing the updating of shared and also the security verification of the data. The updating of the shared key will be done at both sources sensing device side as well as at Data stream manager side. But the security verification will be done only at DSM side.

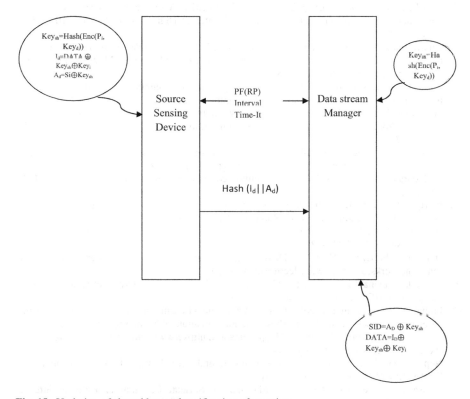

Fig. 18 Updating of shared key and verification of security

6   Conclusion

In this chapter, we discussed two security algorithms. The first algorithm is used to provide the security to the vehicular data. In this methodology, the vehicular nodes and slink nodes need to register at big data center. In this methodology, single sign-on algorithm was used for login to the big data center. Symmetric key cryptography was used in this methodology. The second algorithm is for providing the security to the Sensor data. Here the dynamic prime number based security scheme was used for proving security to the big data. This prime number generation will be done at both sources as well as at big data center side.

References

1. Guo L, Dong M, Ota K, Li Q, Ye T, Wu J, Li J (2016) A secure mechanism for big data collection in large scale internet of vehicle. IEEE Internet Things J, https://doi.org/10.1109/ji ot.2017.2686451
2. Guerrero-ibanez JA, Zeadally S, Castillo JC (2015) Integration challenges of intelligent transportation systems with connected vehicle, cloud computing, and internet of things technologies. IEEE Wirel Commun 22(6):122–128
3. Puthal D, Nepal S, Ranjan R, Chen J (2017) A dynamic prime number based efficient security mechanism for big sensing data streams. J Comput Syst Sci, Science Direct 22–42
4. Walravens C, Dehaene W (2012) Design of a low-energy data processing architecture for WSN nodes. In: Proceedings of the conference on design, automation and test in Europe, March 2012, pp 570–573
5. Walravens C, Dehaene W (2014) Low-power digital signal processor architecture for wireless sensor nodes. IEEE Trans Very Large Scale Integr (VLSI) Syst 22(2):313–321
6. Kaddoura I, Abdul-Nabi S (2012) On formula to compute primes and the nth prime. Appl Math Sci 6(76):3751–3757
7. Perrig A, Szewczyk R, Tygar J, Wen V, Culler DE (2001) SPINS: security protocols for sensor networks. In: Proceedings of ACM MobiCom'01, 2001, pp 189–199
8. Akyildiz I, Su W, Sankarasubramaniam Y, Cayirci E (2002) Wireless sensor networks: a survey. Comput Netw 38(4):393–422
9. Hempstead M, Lyons MJ, Brooks D, Wei G (2008) Survey of hardware systems for wireless sensor networks. J Low Power Electron 4(1):11–20
10. Burke J, McDonald J, Austin T (2000) Architectural support for fast symmetric-key cryptography. ACM SIGOPS Oper Syst Rev 34(5):178–189
11. Puthal D (2012) Secure data collection and critical data transmission technique in mobile sink wireless sensor networks. M. Tech thesis, National Institute of Technology, Rourkela
12. TCG Trusted Platform Module (TPM) specification, https://www.trustedcomputinggroup.org/ specs/tpm/. Accessed on 04 Aug 2014
13. Nepal S, Zic J, Liu D, Jang J (2011) A mobile and portable trusted computing platform. EURASIP J Wirel Commun Netw 2011(1):1–19
14. Gulisano V, Jimenez-Peris R, Patino-Martinez M, Soriente C, Valduriez P (2012) Stream cloud: an elastic and scalable data streaming system. IEEE Trans Parallel Distrib Syst 23(12):2351–2365
15. Guo L, Wu J, Xia Z, Li J (2015) Proposed security mechanism for XMPP-based communications of ISO/IEC/IEEE 21451 sensor networks. IEEE Sens J 15(5):2577–2586
16. Aalm KM, Saini M, Saddik AE (2015) Toward social internet of vehicles concept, architecture, and applications. IEEE Access 3:343–357

17. Cecchinel C, Jimenez M, Mosser S, Riveill M (2014) An architecture to support the collection of big data in the internet of things. In: Proceedings of the IEEE 10th World Congress on services, Anchorage, June 2014, pp 442–449
18. Su Z, Xu Q, Qi Q (2016) Big data in mobile social networks: a QoE-oriented framework. IEEE Netw 30(1):52–57
19. Guo L, Dong M, Ota K, Jun W, Li J (2015) Event-oriented dynamic security service for demand response in smart grid employing mobile networks. China Commun 12(12):63–75
20. Tracey D, Sreenan C (2013) A holistic architecture for the internet of things, sensing services and big data. In: Proceedings of the 13th IEEE/ACM international symposium on cluster, cloud, and grid computing, Delft, May 2013, pp 546–553
21. Salahuddin MA, Al-Fuqaha A, Guizani M (2015) Software-defined networking for RSU clouds in support of the internet of vehicles. IEEE Internet Things J 2(2):133–144
22. Li H, Lu R, Zhou L, Yang B, (Sherman) Shen X (2014) An efficient Merkle tree based authentication scheme for smart grid. IEEE Syst J 8(2):655–663
23. Li H, Lin X, Yang H, Liang X, Lu R, (Sherman) Shen X (2014) EPPDR: an efficient privacy-preserving demand response scheme with adaptive key evolution in smart grid. IEEE Trans Dependable Secure Comput 25(8):2053–2064
24. Li H, Yang Y, Luan TH, Liang X, Zhou L, (Sherman) Shen X (2015) Enabling fine-grained multi-keyword search supporting classified sub-dictionaries over encrypted cloud data. IEEE Trans Dependable Secure Comput
25. Liu C, Zhang X, Liu Z, Yang Y, Ranjan R, Georgakopoulos D, Chen J (2013) An iterative hierarchical key exchange scheme for secure scheduling of big data applications in cloud computing. In: Proceedings of the 12th IEEE international conference on trust, security and privacy in computing and communications, Melbourne, July 2013, pp 10–16
26. Adluru P, Datla SS, Zhang X (2015) Hadoop eco system for big data security and privacy. In Proceedings of the 2015 IEEE Long Island Systems, Applications and Technology Conference (LISAT), Farmingdale, May 2015, pp 1–6
27. Jam MR, Khanli LM, Akbari MK, Javan MS (2014) A survey on security of Hadoop. In: Proceedings of the 4th International Conference on Computer and Knowledge Engineering (ICCKE), Mashhad, Oct 2014, pp 716–721
28. Xu L, Jiang C, Wang J, Yuan J, Ren Y (2014) Information security in big data privacy and data mining. IEEE Access 2:1149–1176
29. Wang H, Qin B, Wu Q, Xu L, Ferrer JD (2015) TPP: traceable privacy-preserving communication and precise reward for vehicle-to-grid networks in smart grids. IEEE Trans Inf Forensics Secur 10(11):2340–2351
30. Soares J, Borges N, Canizes B, Vale Z (2015) Probabilistic estimation of the state of electric vehicles for smart grid applications in big data context. In: 2015 IEEE Power & Energy Society General Meeting, Denver, July 2015, pp 1–5
31. Mershad K, Artail H (2013) A framework for secure and efficient data acquisition in vehicular ad hoc networks. IEEE Trans Veh Technol 62(2):536–551
32. Gulisano V, Jimenez-Peris R, Patino-Martinez M, Valduriez P (2010) Streamcloud: a large scale data streaming system. In: Proceedings of 30th International Conference on Distributed Computing Systems, ICDCS, 2010, pp 126–137
33. Arasu A, Babcock B, Babu S, Datar M, Ito K, Nishizawa I, Rosenstein J, Widom J (2003) STREAM: the stanford stream data manager (demonstration description). In: Proceedings of the 2003 ACM SIGMOD International Conference on Management of Data, 2003, p 665
34. Carney D, Çetintemel U, Cherniack M, Convey C, Lee S, Seidman G, Stonebraker M, Tatbul N, Zdonik S (2002) Monitoring streams: a new class of data management applications. In: Proceedings of the international conference on very large data bases, 2002, pp 215–226
35. Abadi DJ, Carney D, Çetintemel U, Cherniack M, Convey C, Lee S, Stonebraker M, Tatbul N, Zdonik S (2003) Aurora: a new model and architecture for data stream management. VLDB J 12(2):120–139
36. Chandrasekaran S, Cooper O, Deshpande A, Franklin M, Hellerstein JM, Hong W, Krishnamurthy S, Madden SR, Reiss F, Shah MA (2003) TelegraphCQ: continuous dataflow processing.

In: Proceedings of ACM SIGMOD International Conference on Management of Data, 2003, p 668

37. Tatbul N, Çetintemel U, Zdonik SB (2007) Staying fit: efficient load shedding techniques for distributed stream processing. In: Proceedings of international conference on Very Large Data Bases, VLDB, 2007, pp 159–170

38. Jin M, Zhou X, Luo E, Qing X (2015) Industrial-QoS-oriented remote wireless communication protocol for the internet of construction vehicles. IEEE Trans Industr Electron 62(11):7103–7113

39. Puthal D, Sahoo B (2012) Secure data collection & critical data transmission in mobile sink WSN. In: Secure and energy efficient data collection technique. LAP Lambert Academic Pubilishing, Germany, ISBN978-3-659-16846-8

40. Kumar N, Rodriguesand JJPC, Chilamkurti N (2014) Bayesian coalition game as-a-service for content distribution in internet of vehicles. IEEE Internet Things J 1(6):554–555

41. Fu J, Chen Z, Sun R, Yang B (2014) Reservation based optimal parking lot recommendation model in internet of vehicle environment. China Commun 11(6):38–48

42. Cheng J, Cheng J, Zhou M, Liu F, Gao S, Liu C (2015) Routing in internet of vehicles a review. IEEE Trans Intell Transp Syst 16(5):2339–2351

43. Dua A, Kumar N, Bawa S (2014) A systematic review on routing protocols for vehicular ad hoc networks. Veh Commun 1(1):33–52

44. Li B, Zhao C, Zhang H, Sun X (2013) Characterization on clustered propagations of UWB sensors in vehicle cabin: measurement, modeling and evaluation. IEEE Sens J 13(4):1288–1300

45. Kumar N, Misra S, Rodrigues J, Obaidat MS (2015) Coalition games for spatio-temporal big data in internet of vehicles environment: a comparative analysis. IEEE Internet Things J 2(4):310–320

46. Zhou Y, Chen S, Zhou Y, Chen M (2015) Privacy-preserving multi-point traffic volume measurement through vehicle-to-infrastructure communications. IEEE Trans Veh Technol 64(12):5619–5630

47. Wu Q, Ferrer JD, Nicolas ÚG (2010) Balanced trustworthiness, safety, and privacy in vehicle-to-vehicle communications. IEEE Trans Veh Technol 59(2):559–573

48. Li H, Liu D, Dai Y, Luan TH (2015) Engineering searchable encryption of mobile cloud networks: when QoE meets QoP. IEEE Wirel Commun 22(4):74–80

49. Cárdenas AA, Manadhata PK, Rajan SP (2013) Big data analytics for security. IEEE Secur Priv 11(6):74–76

IoT for Ubiquitous Learning Applications: Current Trends and Future Prospects

Salsabeel Shapsough and Imran A. Zualkernan

Abstract Concept of ubiquitous learning is not new, and over the years, ubiquitous learning applications have migrated from the web to mobile-based applications with an increased use of onboard sensors. The current surge in Internet of Things (IoT) has the potential to enable another shift towards a new class of ubiquitous learning applications that rely on cheap sensors, edge devices, and IoT middleware and protocols to enact unique and innovative ubiquitous learning designs. This chapter will first present a survey of key ubiquitous learning applications that use sensors and edge devices in a meaningful manner, showing that such applications are typically constructed in an ad hoc manner, and mostly use web-based protocols like the HTTP. After the survey, this chapter will present a typology of current ubiquitous learning applications and map this typology to the various components of the IoT stack. Representative ubiquitous learning systems in the areas of language and social sciences learning, science and technology learning, and domain independent learning will be considered. Thirdly, based on recent trends in IoT technologies, new architectures and future, prospects for IoT-based ubiquitous learning applications will be explored.

1 Introduction

Ubiquitous Learning is defined in literature as learning anything, anywhere, anytime. A ubiquitous learning application is designed to eliminate the constrictions of a typical classrooms by presenting customizable learning content over platforms familiar to learners [1]. The recent advancements in internet computing technologies, however, have morphed the term into a wide umbrella under which fall various applications that incorporate sensing and communication technologies for a situated, context-aware learning environment [2]. Using interactive elements such as RFID tags, QR-codes, and smart sensors, learning can take place outside classrooms

S. Shapsough (✉) · I. A. Zualkernan
Department of Computer Science and Engineering, American University of Sharjah, Sharjah, United Arab Emirates
e-mail: g00041619@aus.edu

© Springer Nature Switzerland AG 2019   53
N. Jeyanthi et al. (eds.), *Ubiquitous Computing and Computing Security of IoT*,
Studies in Big Data 47, https://doi.org/10.1007/978-3-030-01566-4_3

and in authentic settings. Over the years, dozens of environments and mobile applications have been developed to implement the concept of ubiquity in learning. In [3], Njoku analyzes 135 studies on ubiquitous learning environments conducted in the year 2015-106, retrieved from the seven top journals on education and learning technologies. Among the main research questions were: "which subjects do ubiquitous learning studies focus on the most?" and "at which academic level are studies more likely to participate in these studies?" Analyzing the 135 articles showed that most studies are aimed at undergraduate students enrolled in science courses, while the least considered subjects were law and mathematics, and the least common test group was postgraduate students. Furthermore, the analysis found a 100% increase in the number of publications ubiquitous learning studies and environments per year, compared to the number between the years 2010 and 2013 [4] which, at the time, suggested a steadily-growing interest in the field over the past decade. Most of those systems allow communication with Learning Management Systems, providing an opportunity to scale such systems to incorporate learners from different schools.

2 Previous Work on Ubiquitous Learning Applications

Ubiquitous learning environments depend mainly and heavily on two types of technologies: asset tracking and sensing technologies to facilitate context incorporation, and user applications. This led to such environments evolving to reflect the most recent trends in both fields.

2.1 Early Attempts at Ubiquitous Learning Environments

Ubiquitous technologies have repeatedly been used to aid students learning languages, history, and science by creating interactive environments. Such environments place learners in situations where they either interact with ubiquitous technologies to access information, or use their knowledge to engage in activities and complete tasks that involve using ubiquitous technologies. Early ubiquitous platforms were designed mostly for language and culture studies. The platforms were designed mainly for PDA users, and employed features such as cameras, microphones, and wireless communication. Language platforms presented learners with tasks that require using new vocabulary to identify locations and objects, and correct grammar to communicate with locals and peers [5–7]. Saito et al. presents a study on basic support for ubiquitous learning (BSUL) environments; light weight and simplicity, adaptiveness and customization, support for various learning styles, and support for collaborative learning [5]. Such systems have often been implemented using internet-connected PDAs and personal computers in coordinance with existing Learning Management Systems (LMS); thus, allowing students to seamlessly learning experience where they have access to any information they need, whenever they need it. To conduct the study, a ubiquitous learning software was developed implementing the activities seen as most important in classrooms; including: report submission and attendance taking, in addition to student feedback and assessment response. The system was used in multiple case studies, most notably a Language-learning outside the classroom with handhelds (LOCH) case study [6] where students enrolled in a Japanese language course were asked to walk around town and perform certain tasks, aided by their PDAs. For example, students were asked to engage in conversations with locals to acquire certain information such as prices of different products and different outlets' opening hours. Another type of tasks required them to conduct interviews with school faculty in Japanese, and use their PDA to take a photo of the interviewee and note down important information. The PDAs were connected to the internet; meaning that the instructor will be able to track students on a map and update them with new tasks and/or feedback. Chen et al. introduces a personalized context-aware ubiquitous learning system (PCULS) that helps students learn English by focusing on vocabulary [7]. By taking the learner's vocabulary level, learning time, and their location—acquired using wireless network positioning—the environment is able to adapt and customize the content in order to provide a better learning experience.

In culture study courses, learners on a fieldtrip can learn about the history and culture of a location by completing tasks that required them to walk to certain locations and answer questions related to the history of said location. Museums, for example, often resort to electronic trails in order to provide guests with an educational experience [8]. A museum trail is typically a paper, brochure, or pamphlet that guides guests through different sections, providing brief historical facts about each object. Electronic trails provide the same information but in different forms, including audio and video. Electronic trails also allow museum visitors to add feedback and reflections in the different forms. This is particularly useful for visitors who wish to access the information later on; to prepare a presentation about their visit, for example. In [9], students from 10 primary schools learning about the history of Amsterdam were asked to walk around the city, observing various monuments and learning about historical characters and stories hidden within their walls. Each group of students was divided into a city team (CT) and a headquarter team (HQT). The CT students will then walk around the city and use UMTS/GPS phones to perform certain tasks and share new findings with the HQT students, while HQT students provided directions using a modern map and a medieval map of the city; with the end goal being a full narrative that connects various historical elements of the city. The project builds on the concept of "storification", where students are more likely to understand and remember historical characters and events when they are woven into a narrative. Being in the same vicinity as said elements adds a dimension of reality, helping students draw a full picture of events that took place around the city. Learners using these platforms to request, contribute, and exchange text, audio, and video-based information that aid them in completing different tasks. Location services such as GPS and WiFi localization were often used to either customize learning material, or confirm a learner has completed a certain task by arriving at a particular location. The platforms also often made use of wireless communication and messaging protocols to facilitate peer-to-peer communication.

2.2 The Introduction of Passive Sensors to Ubiquitous Learning

With subjects like natural science and engineering courses, on the other hand, the focus is on the learners' ability to identify and interact with physical objects. Before the incorporation of technologies such as QR codes and barcodes, physical objects served as "observables". The learners' inputs about them in the form of remarks and questions gave the objects presence in the system. The availability of QR-codes, barcodes, and RFID tags at a low cost changed that by giving physical objects and locations a real presence in the virtual system. Learning material and tasks could be tied to certain locations and objects using code, and learners can interact with them by scanning them [10–15].

The authors in [10] present the results of a study that investigates the effect of context-aware ubiquitous learning environments on the performance of English learners. In the study, students are divided into two groups; a control group, and an experimental group. Both groups were asked to perform 3 activities: practice listening and speaking during their free time, participate in activities during class time, and perform a story relay. The control group was provided with classical learning aid such as printed materials, audio CDs, and voice recorders. The experimental group, on the other hand, was provided with HELLO (Handheld English Language Learning Organization); an application that runs on their mobile phones and presents each of the tasks in the form of a game that they can play outdoors. Context-awareness is implemented using area-specific QR codes; which students use their phones to scan and download different learning material depending on their location. The authors of [11] present the design and evaluation of a u-learning application designed to ease the process of editing ubiquitous learning material. In their work, Chin et al. present a system that consists mainly of a mobile application and a main server. Students walk around an area where certain elements have been labeled with QR-codes. A student can scan a QR-code and download learning content that has been previously edited and uploaded by their teacher. Students are then asked to use this information to perform different tasks. The system was used in a study where first year college students learnt about the Taiwanese Cultural Heritage; in which QR-code were placed on different artifacts such as statues and old journals. Not only did students report a higher level of enjoyment compared to typical learning methods, but they were able to achieve higher scores. Similarly, [12] presents a game that combines the use of QR code with the concept of prompt-based learning. In this game, students studying 5th grade-level natural science are assigned unique locations in a garden which serves as a home for different kinds of plants. Students move around the garden based on rolls of dice, and earn points by completing tasks where they are asked to locate a certain plant and answer questions about its characteristics using material they download by scanning QR codes. Scanning the QR also serves as a way to locate the student and identify their context, which is then used to form the next task. Hung et al. describes a similar system in [13] under the title Context-aware Reflection Prompt System (CRPS), with the main difference being the prompt and response type. This is motivated by studies that claim students enjoy a better learning experience when prompts arc in video format rather than text. Not only are videos believed to be more engaging, but they fit better into real-life narratives. Liu et al. presents an environment of ubiquitous learning with educational resources (EULER) designed to aid students study natural sciences by going to outdoor locations [14]. Students, for example, can learn about animals by walking around a zoo and accessing context-specific learning material at different zones. This is made possible by placing RFID tags at different areas and cages; where the student will use their devices to scan the RFID tag and download available information uploaded by their teacher. Groups of students then combine their findings into a report to be reviewed by their teacher. Another, similar system that uses QR codes and NFC objects to teach students about hardware components of a computer is presented in [15]. One thing that is emphasized in the latter but never mentioned by the previously is that many of these systems are what is now considered the simplest form of an Internet of Things application. RFID, QR codes and NFC are the some of the easiest and most common ways to turn everyday objects into "things". The reoccurring pattern among those systems is learners using PDAs or smartphones to scan codes, download learning content, and communicate with peers using HTTP or GSM over a wireless or a cellular network. In [15], Gomez et al. show how a QR code/NFC-based ubiquitous learning system can fit into the classical IoT paradigm; which consists of three layers: things, communication technologies, and clients. In this case, QR codes and NFC objects are things, websockets over WiFi is the communication layer, and a mobile application allows users to interact with the system at the client level. The relation between IoT and ubiquitous learning environments was also recognized by [16, 17]. In [16] where an IoT-based ubiquitous learning system incorporates NFC and QR technologies to offer a Teaching by Example and Learning by Doing (TELD) experience for students studying perishable products auction. In this application, products offered for auction are tagged with QR codes and NFC tags. Students playing as either bidders or auctioneers use a mobile application to scan the objects to perform actions such as offer an object for auction, set initial price, bid on objects, and mark objects as sold. Students can also access information related to the offered objects, as well as market dynamics. The auction experience is carried out over a wireless network via WiFi or Bluetooth. Finally, in [17] students study Logistics and Supply Chain Management but immersing themselves in three TELD smart environments. A learning factory, learning distribution center, and learning retail shop are made smart by equipping objects, areas, and checkpoints with RFID tags and readings. The RFID hardware aspect is used in addition to wireless communication and cloud computing software technologies to simulate a real-life supply chain and provide hands-on training.

This all leads to the question of "how much can IoT technologies offer to the field of ubiquitous learning?"

2.3   IoT-Assisted Ubiquitous Learning

With IoT-based technologies gaining momentum the recent decade or so, cheap sensor technologies became more readily available than ever; and subsequently found their way into u-learning. There are several examples of ubiquitous environments where wearable sensors and simple microcontrollers were employed to transform the learning experience into a game of hide-and-seek; adding an element of healthy physical activity to academics [18, 19]. Cheap active and passive sensors can be also be used to acquire quantitative readings such as temperature and humidity in an area [18, 19], or provide input on the state and dimensions of an object [19]. Prete-a-apprendre [18] is a u-learning system that incorporates the concept of context-aware ubiquitous learning into primary school mathematics, but adds an element of physical activity to it. In this system, students are asked to come up with three true-or-false math questions each at the beginning of the game. After the questions have been collected, the game begins and a random question is assigned to one student each round. The assignment is done using wearable, zigbee-equipped Lilypad microcontrollers sewn into each student's shirt. The shirt also has two pressure sensor patches; one for "true", and another for "false". In each round, the shirt of the student that has been assigned a question lights up, and other students must try to catch them and answer the question using the pressure sensors. Hoodies and Barrels [19] is another, similar system where children play hide and seek based on mathematical constrains set either by their teacher or by themselves. The system also incorporates wearable technology that is used to send children rules for the type of place where they can hide; e.g. behind a barrel that is 1 m^2 in volume. The child has to formulate the rules themselves and then send it to their peers; who then use that rule to find them. In terms of mathematical knowledge, while Prete-a-apprendre focuses on theory and students doing mathematical operations on the fly, Hoodies and Barrels is more concerned with students' perception of concepts such as shapes and volumes. The latter trait is also often the main focus of u-learning systems developed for science courses such as biology and geology. OBSY [20] is an excellent example of systems that implement context-aware u-learning which emphasize the students' perception of and interaction with their physical surroundings. The developers make use of old PC tablets that were distributed to schools in rural Northern Thailand as a part of the One Tablet PC per Child policy. Each student is offered a Raspberry Pi-based sensor node equipped with a camera and a group on sensors that they can interact with via an application deployed on their tablet. The system aims to connect students to their environment by offering them the chance to associate their own experience with a space with quantitative values acquired by sensors. Using the camera, students are also able to monitor elements in their environment—for example, the growth of a small plant—in order to add a realistic experiential value to is usually taught through textbooks. The student can move around campus, position their personal node (OBSY) anywhere they wish, and take sensor readings and footage to be used in answering different questions. Another example of context-aware ubiquitous learning with inquiry-based learning is Archi-Pods [21]. In this system, students enrolled in an

Architecture Principles university-level course are provided with web-based mobile applications that interact with sensor nodes distributed around campus. The nodes are equipped with various sensors including temperature, motion, and illumination. Students are asked to walk around campus, interact with sensor nodes to acquire sensor readings, and then incorporate this information into constructing or answering interesting questions about a particular space. In one task, for example, students are asked to locate an architectural element such as an arc. Once a student has located an arc, they can take a photo using their mobile phones and use it to answer the question. In another task, students are asked to find illumination level in a space, and relate it to how the space is being used. The system also allows students to rate questions, which in turn works as a mechanism for improving their question posing skills and elevating critical thinking.

In such systems, sensing nodes would be spread out over a particular location such as a campus or a garden, and learners would be required to walk around the area, communicating with different nodes and completing location-unique tasks. Micro-controllers like Particle's Photon and Electron [22], Arduino [23], and Adafruit's Feather series [24] are used to integrate sensor readings. More sophisticated sensor nodes are built using microcomputers like Raspberry Pi [25], C.H.I.P [26], Beagle-Board [27], ODROID [28], Samsung's ARTIK series [29], and Intel's Edison [30]. In addition to sensing, microcontrollers and microcomputers provide varying levels of networking, messaging, and application hosting. One group of researchers suggests using learning robots as the sensing node element [31]. Chen et al. propose the design of a subject-independent Ubiquitous Open-structured Neo-tech Edutainment (u-ONE) architecture which combines learning robots, sensing technology, mobile computing, and wireless networks [31] to support children's learning. The framework is designed based on the concepts of experiential and gamified learning, and can be used in instruction, collaborative learning, and adaptive self-learning. A total of five hardware component types are defined in an implementation of the system: a wireless network, an output element, a mobile computing element, perception elements like sensor and RFID, and a learning robot. The inclusion of all five elements is claimed to be essential, although the choice of product for each category and how they are integrated into the system is left to the developer. This is meant to provide flexibility in terms of customization and cost; as implementation may differ depending on the subject being taught, and more importantly-on the availability and affordability of different products. Each student is provided with one or more perception elements to help them interact with their surroundings, and a learning robot that serves as their access point to the system. On the other side of the classroom, the instructor is provided with a mobile computing device capable of collecting and processing student performance data, and any sort of output screen to display the results. For the most part, the framework mimics all of the systems discussed earlier; with the exception of the learning robots. In addition to adding a touch of fun into the learning environments, they can provide real-time one-on-one interaction with every student and real-time feedback; thus, reducing the pressure on the instructor who will otherwise have to divide their time, attention, and energy among all students. Three scenarios have been suggested for how the system may be used. In case of an instruction-based class period, instruction material will be distributed to the learning robots, and students will have equal chances to participate during the lecture. The same learning robot can also be used to communicate with other students in a collaborative learning session, or provide adaptive learning material in the case of self-learning. Wireless communication between different stations is most crucial in the first and second scenarios, but is not needed for self-learning; as most learning material is pre-downloaded into the station at the beginning of the session.

Other systems went as far as employing sensors embedded in learners' smartphones [32, 33], and smart technology like Google Glass [34] as the sensing element. This provided learners with access to environmental information on the go, allowing them to conduct comparisons between their personal experience and real quantitative information, and document their experience as they go. In [33], Shapsough et al. present the design and evaluation of an assessment system where students use smartphones to access and answer assessments set by their teachers. The use of sensor data in this particular system differs from other discussed systems because the readings are not involved in the students' answers; but are relevant to understanding the context in the student was while answering the assessments. However, only few application-level modifications will be needed in order to change the system so that it allows students to access the sensor information and incorporate them in their answers. WeSPOT (Working Environment with Social, Personal and Open Technologies for Inquiry Based Learning) [32] is another example that implements this category of learning systems. It is an inquiry-based learning system based on the Personal Learning Environment (PLE) paradigm-a learner-centric environment which provides tools and services for students to design their own learning experience. WeSPOT aims to patch the gap between different inquiry tools and services, students' profiles, and the curricular con-text in order to create a cohesive learning environment tailored to the students' needs. The environment offers students a game-like experience that is meant to mimic the sort of interaction that takes place in social media. This, the developers believe, motivates the students to take advantage of the system's capabilities as they feel a sense of accomplishment earning points in form of badges as they reach certain milestones. Students can access the system through their mobile application that they can download on their smartphones and proceed to pose and answer questions that fall under tasks specified by the instructor. The system also supports a type of tasks where students are asked to collect visual and audio information from their environment through their smartphones, which incorporates the element of personal experience into the learning process. Google Glass Personal Inquiry Manager (Glass PIM or GPIM) [35] is a Glassware (Google Glass application) that was introduced in order to elevate the WeSPOT experience. GPIM focuses on the experimental aspect of the environment as students can now incorporate images, audio, and video taken by their Google Glass unit into their inquiries. The incorporation of Google Glass is meant to provide smoother flow of information between the students' physical surroundings and the learning system, and makes it easier for the student to keep track of their peers' contributions while working to complete their tasks, instead of having to constantly shift their attention between the application and their surroundings.

3 The Generic IoT Architecture

There are several published works that compared available e-learning and ubiquitous learning platforms. In addition to the ones previously discussed in [3] and [4], the work done [36, 37] present comparisons between different platforms in literature. In both publications, the authors survey a number of e-learning and u-learning environments, and compare them in terms of technical support, and case study dimensions. Such comparisons are useful to acquire a general picture of the technologies involved in u-learning environments. However, in order to bring the field of u-learning into the Internet of Things (IoT) paradigm, it is more sensible to examine existing systems in the light of said paradigm. Different authors, developers, and stakeholders present different versions of the IoT generic architecture; which often vary in the number of layers defined and the classification of technologies into the layers. This arises from the fact that people from different backgrounds are taking part in the IoT revolution [38]. A corporation, for example, will define a paradigm that more based on supply chain and business operations, while a telecommunication institute will focus more on the communication technologies and topologies. A generic architecture is derived from ones proposed in [38–42]. The architecture consists of four layers: perception, network, middleware, application (Fig. 1).

The Perception Layer (PL), sometimes referred to as the Device Layer- is provides interfaces between the system and the physical world, mainly identifying objects and context, and acquiring data. Technologies that fall under this layer are sensors, QR codes, RFID, smart meters, and others. The next layer is the Network Layer (NL), which provides communication media between PL devices and upper layers in the architecture. The layer covers small-scale and wide-scale networks, and employs wireless communication technologies such as Wi-Fi, Bluetooth, and others. On top of the NL is the Middleware Layer (ML), which provides management services as

Fig. 1 Internet of Things architecture

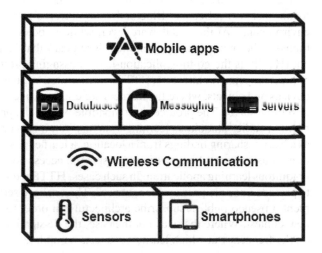

well as databases, processing services, messaging brokers, and others. The layer is responsible for handling the heterogeneity of the devices in the IoT system and providing a seamless exchange of information. Following the ML is the Application Layer (AL) which provides interfaces between the users and the IoT system in the form of mobile applications, websites, reports, and others. A comparison of the discussed ubiquitous environments can then be conducted based on how they fit into IoT paradigm (Table 1). Some of the systems are custom-made; the developer is given the responsibility of choosing the product/technology for each element, and then integrate them together. This is highly inefficient considering that most systems, in essence, have the same requirements and are designed to achieve the same goal. A unified architecture for this class of systems will massively reduce the design and implementation time required to produce a system instance. Other works such as [11, 31–34], on the other hand, attempt to implement said unified architecture.

4 Ubiquitous Learning Applications Mapped to the IoT Architecture

Table 1 demonstrates the main components of each reviewed ubiquitous learning environment as they fit into the IoT architecture's four layers. From analyzing the components of each application, one can notice a few interesting trends. At the perception layer, the sensing element in earlier systems was features that were already included in PDAs such as cameras and microphones. As sensing technologies became more commercialized, ubiquitous learning environments often adopted such technologies and edge devices increased in sophistication. The same trend is also present at the application layer, where front-end applications moved from being designed for PCs and browsers, to adopt latest technologies in mobile and cross-platform apps. At the network layer, WiFi is the most common wireless protocol, with lighter protocols such as ZigBee and Bluetooth being offered as additional options in several environments. At the middleware layer, while middleware and databases are often hinted at but not explicitly identified, it appears that Hyper Text Terminal Protocol (HTTP) is the go-to application-level messaging protocol. Natively based on a request-response architecture, HTTP is well-suited for interactions between learners' devices and objects, where the learners are clients requesting resources from a local or remote server. The protocol is compatible with web applications, which were the norm in earlier ubiquitous learning systems. Yet, it is peer-to-peer transactions such as a learner sharing findings from a location, a learner posing a context-specific question, or assessments being assigned to specific access points are at the heart of every ubiquitous learning application. In such cases, HTTP requires additional middleware to provide a pseudo peer-to-peer network. The backend servers also needed to implement a pseudo publish/subscribe architecture in order to keep learners updated on new contest. Where that was not the case, the dissipation of information depended on the frequency at which learners requests an update.

Table 1 Ubiquitous learning platforms as they fit into the IoT architecture

System	Year	Perception layer		Network layer	Middleware layer			Application layer	Data format
		Edge device	Sensing element		Messaging	Database	Middleware		
PCULS [7]	2010	Phone	WLAN localization	WiFi	SMS, HTTP	SQL	N/S	Web app	T
Museum [8]	2010	Phone	Camera, mic	WiFi	HTTP	N/S	N/S	Web app	T, I, A, V
Amsterdam [9]	2009	Phone	GPS, camera, mic	UMTS	SMS, MMS	–	N/S	Browser app	T, I, A, V
LOCH [6]	2008	Phone	GPS, camera, mic	PHS	HTTP	N/S	N/S	J2EE web app	T, I, A
Natural science [12]	2015	Phone	QR	WiFi	HTTP	N/S	N/S	Android app	T, V
CRPS [13]	2014	Phone	QR	WiFi	HTTP	N/S	N/S	Android app	T
EULER [14]	2009	RFID reader, PDA	RFID, Camera	WiFi	HTTP	SQL	N/S	Web app	T, I, A, V
IoT-based UL [15]	2009	Phone	NFC, QR	WiFi, 3G	WebSocket	N/S	N/S	X-platform app	T, I, A, V
Dutch Auction [16]	2017	Phone	NFC, QR	WiFi, Bluetooth	N/S	N/S	N/S	Android app, PC app	T
LSCM [17]	2017	Phone	RFID	WiFi, Bluetooth	HTTP	N/S	N/S	Web app	T
OBSY [20]	2015	RPi	Sensors, camera	WiFi	HTTP	–	N/S	Web app	T, I

(continued)

Table 1 (continued)

System	Year	Perception layer		Network layer	Middleware layer			Application layer	Data format
		Edge device	Sensing element		Messaging	Database	Middleware		
Prete-a-apprendre [18]	2010	Arduino	Sensors	WiFi, ZigBee	HTTP, ZigBee	–	ZUL	PC app	T
Hoodies and Barrels [19]	2011	Arduino, RFID reader	Sensors, RFID	WiFi, ZigBee	ZigBee	–	ZUL	PC app	T
HELLO [10]	2010	Phone	QR	WiFi/WCDMA	HTTP	SQL	N/S	Web app	T, I, A, V
QR-Lumps [11]	2015	Phone	QR	WiFi	HTTP	N/S	N/S	Android app, PC app	T, I, A, V
Context-aware assessment [33]	2016	Phone	Sensors, GPS	WiFi	MQTT	CDB	Custom	X-platform app	T
WeSpot [43]	2014	Phone	Camera, mic, sensors	WiFi	HTTP	My SQL	Elgg (PHP)	Web app, Android app	T, I, A, V
GPIM [34]	2014	Google Glass	Sensors	WiFi	HTTP	My SQL	Elgg (PHP)	Google Glassware	T, I, A, V
u-ONE [31]	2010	Robot, RFID, PDA, iPod, OLPC	Barcode, RFID, QR, E-Pen	BLE, WiFi, ZigBee, Group Net	BLE, HTTP, ZigBee	N/S	N/S	PC app, Android app	T

T text; *V* video; *I* image; *A* audio

5   Future Prospects and Recommendations

Beside cheap sensing and edge computing technologies, the IoT trend generated a flux of new network and software technologies in the form of messaging protocols, middleware, and application ecosystems. It is common for platforms to use new network protocols which are less resource-demanding such as Bluetooth Low Energy, ZigBee, Thread, etc. This is driven by the IoT's common dependence on wide-scale organisms of heterogeneous constrained devices. The same goes for software developed for the middleware and application layers. Those technologies are often designed to be lightweight, scalable, and low on resources consumption. Because ubiquitous learning systems fit so smoothly into the IoT paradigm, replacing classic software and protocols with IoT technologies is expected to improve the performance of such systems.

At the backend as a part of a middleware, and on edge nodes, ubiquitous systems can benefit from using application environments such as NodeJS. Besides being lightweight and scalable, NodeJS is supported on most common platforms, and is based on JavaScript, which is becoming the lingua franca of smart systems. The environment can also be easily integrated with noSQL databases. In IoT systems, noSQL databases are favorable because data is rarely structured and is variable from one implementation to another. Document-based noSQL databases such as MongoDB [44] and CouchDB [45] which are document-based allow for a high level of flexibly in terms of data format. They also favor JSON, which fits in with a purely-JavaScript ecosystem, and is supported by all common messaging protocols.

Real-time communication is at the core of any ubiquitous learning platform, and yet, as mentioned earlier, HTTP is often the application-level communication protocol of choice. Several IoT Application-level messaging protocols have been developed as alternative to heavier protocols such as HTTP. This was driven by the peer-to-peer nature of IoT communication where even the simplest IoT application would consist of dozens of distributed devices, each serving as a resource, observing one, or both, with no central server entity. This is unlike HTTP which is too rigid in terms of client and server dynamics compared to IoT environments. Publish/Subscribe, Observer, and even Instant Messaging protocols such as Message Queue Telemetry Transport (MQTT), Advanced Message Queuing Protocol (AMQP), Simple Text Oriented Messaging Protocol (STOMP), Constrained Application Protocol (CoAP), and Extensible Messaging and Presence Protocol (XMPP) are among the top contenders, based on their native architecture. The protocols are also event-based, meaning messaging is initiated by an event taking place at a resource/publisher, rather than observers/subscribers continuously checking for updates. This not only saves power, but also reduces unnecessary traffic. Furthermore, HTTP can prove resource-heavy and inefficient, especially as the system increases in scale, while participating devices become more mobile and constrained. At the same time, most IoT protocols are designed to support this type of messaging at a lower computation, storage, and transport cost. They are also fully supported and easily integrate-able with application environments such as NodeJS at the backend, and native and cross-platform applications at the front end. Going back to ubiquitous learning environments, messaging is more naturally implemented using IoT protocols, where one or more learners or instructors subscribes to receive updates from one or more resources, and are automatically updated with new events by the broker. The protocols offer a reduction in storage and computation cost, giving the system designer more freedom in choosing edge devices that could be constrained in both.

6 Conclusion

This chapter did three things. First, it presented a survey of key ubiquitous learning applications in an order that reflected how, while the concept was remains relatively the same, trends in computing technologies helped such applications evolve to offer more possibilities to developers and users alike. While earlier applications made use of phone cameras and microphones, asset tracking technologies such as QR codes, and cheap sensors and microcontroller made it possible to incorporate more of a learner's context into their learning. Based on those observations, the applications were mapped into a generic Internet of Things (IoT) paradigm, which revealed that ubiquitous learning was a domain which could easily adopt and make use of current trends in IoT. Based on that, the last part of the chapter presented recommendations as to how current trends in IoT computation and communication technologies can benefit such applications, and what are future prospects for IoT-assisted ubiquitous learning applications.

References

1. Yang SJH (2006) Context aware ubiquitous learning environments for peer-to-peer collaborative learning. J Educ Technol Soc 9(1):188–201
2. Hwang G-J, Tsai C-C, Yang SJH (2008) Criteria, strategies and research issues of context-aware ubiquitous learning. J Educ Technol Soc 11(2):81–91
3. Njoku MGC (2016) Trend analysis of mobile and ubiquitous learning: 2014–2015. Int J Mob Learn Organ 10(3):117–128
4. Gunay and Yakin (2014) The status of mobile and ubiquitous learning: a content review of the recent researches. Ubiquitous Learn Int J 6(3):35–45
5. Saito NA, Ogata H, Paredes RGJ, Yano Y, Martin GAS (2005) Supporting classroom activities with the BSUL environment. In: IEEE International Workshop on Wireless and Mobile Technologies in Education (WMTE'05), p 8
6. Ogata H, Saito NA, Rosa G, Paredes J, San Martin GA, Yano Y (2008) Supporting classroom activities with the BSUL system. Educ Technol Soc 11(1):1–16
7. Chen C-M, Li Y-L (2010) Personalised context-aware ubiquitous learning system for supporting effective English vocabulary learning. Interact Learn Environ 18(4):341–364
8. Reynolds R, Walker K, Speight C (2010) Web-based museum trails on PDAs for university-level design students: design and evaluation. Comput Educ 55(3):994–1003
9. Akkerman S, Admiraal W, Huizenga J (2009) Storification in history education: a mobile game in and about medieval Amsterdam. Comput Educ 52(2):449–459

10. Liu T-Y, Chu Y-L (2010) Using ubiquitous games in an English listening and speaking course: impact on learning outcomes and motivation. Comput Educ 55(2):630–643
11. Chin KY, Lee KF, Chen YL (2015) Impact on student motivation by using a QR-based U-learning material production system to create authentic learning experiences. IEEE Trans Learn Technol 8(4):367–382
12. Chen CH, Hwang GJ (2015) Improving learning achievements, motivations and flow with a progressive prompt-based mobile gaming approach. In: 2015 IIAI 4th international congress on advanced applied informatics, 2015, pp 297–302
13. Hung I-C, Yang X-J, Fang W-C, Hwang G-J, Chen N-S (2014) A context-aware video prompt approach to improving students' in-field reflection levels. Comput Educ 70:80–91
14. Liu T-Y, Tan T-H, Chu Y-L (2009) Outdoor natural science learning with an RFID-supported immersive ubiquitous learning environment. J Educ Technol Soc 12(4):161–175
15. Gómez J, Huete JF, Hoyos O, Perez L, Grigori D (2013) Interaction system based on Internet of Things as support for education. Procedia Comput Sci 21:132–139
16. Kong XTR, Chen GW, Huang GQ, Luo H (2017) Ubiquitous auction learning system with TELD (Teaching by Examples and Learning by Doing) approach: a quasi-experimental study. Comput Educ 111:144–157
17. RFID-enabled learning supply chain: a smart pedagogical environment for TELD. Research-Gate. [Online]. Available: https://www.researchgate.net/publication/287481361_RFID-enable d_learning_supply_chain_A_smart_pedagogical_environment_for_TELD. Accessed 17 July 2017
18. Zualkernan IA, Al-Khunaizi N, Najar S, Nour N (2010) Prête-à-apprendre+: towards ubiquitous wearable learning. In: 2010 10th IEEE international conference on advanced learning technologies, pp 740–741
19. Arroyo I, Zualkernan IA, Woolf BP (2011) Hoodies and barrels: using a hide-and-seek ubiquitous game to teach mathematics. In: Advanced learning technologies (ICALT), 2011 11th IEEE international conference on, pp 295–299
20. Putjorn P, Ang CS Farzin D (2015) Learning IoT without the 'I'-educational Internet of Things in a developing context. In: Proceedings of the 2015 workshop on do-it-yourself networking: an interdisciplinary approach, New York, NY, USA, 2015, pp 11–13
21. Shapsough S, Amin S, Ahmed H, Zualkernan I, Mitchell K (2015) ARCHI-PODS: ubiquitous learning technology to teach architectural design principles to architecture students. In: Presented at the INTED2015, 2015, pp 4338–4347
22. Particle [Online]. Available: https://www.particle.io/
23. Arduino—Home
24. Overview| Adafruit Feather 32u4 Bluefruit LE|Adafruit Learning System [Online]. Available: https://learn.adafruit.com/adafruit-feather-32u4-bluefruit-le/overview. Accessed Mar 26 2017
25. Raspberry Pi—Teach, Learn, and Make with Raspberry Pi
26. N. T. Co, "Get C.H.I.P. and C.H.I.P. Pro—the smarter way to build smart things. Next Thing Co. [Online]. Available: https://getchip.com/pages/chip. Accessed Mar 25 2017
27. BeagleBoard.org—black [Online]. Available: https://beagleboard.org/black. Accessed Mar 26 2017
28. ODROID|Hardkernel [Online]. Available: http://www.hardkernel.com/main/main.php. Accessed Mar 26 2017
29. Samsung ARTIK IoT Platform—Samsung ARTIK 1020 IoT Module [Online]. Available: https://www.artik.io/modules/artik-1020/. Accessed Apr 02 2017
30. The Intel® Edison Module|IoT|Intel® Software [Online]. Available: https://software.intel.co m/en-us/iot/hardware/edison. Accessed Mar 25 2017
31. Chen NS, Hung IC, Wei CW, Developing ubiquitous learning system with robots for children's learning. In: 2010 Third IEEE international conference on digital game and intelligent toy enhanced learning, 2010, pp 61–68
32. Bedek MA, Firssova O, Stefanova EP, Prinsen F, Chaimala F (2015) User-driven development of an inquiry-based learning platform: qualitative formative evaluations in weSPOT. Interact Design Architecture(s) 23:122–139

33. Shapsough S, Shapsough SE, Hassan M, Zualkernan I (2016) IoT technologies to enhance precision and response time of mobile-based educational assessments. In: Presented at the 2016 international conference on computational science and computational intelligenc, Las Vegas, USA
34. Suarez A, Ternier S, Kalz M, Specht M, GPIM: Google glassware for inquiry-based learning. In: Open learning and teaching in educational communities. Springer, pp 530–533
35. Suarez A, Ternier S, Kalz M, Specht M, GPIM: Google glassware for inquiry-based learning. Interact Design Architecture(s) 24:100–110
36. Abdullah F, Ward R (2016) Developing a general extended technology acceptance model for e-learning (GETAMEL) by analysing commonly used external factors. Comput Hum Behav 56:238–256
37. Chang H-Y et al (2015) A review of features of technology-supported learning environments based on participants' perceptions. Comput Hum Behav 53:223–237
38. Atzori L, Iera A, Morabito G (2010) The internet of things: a survey. Comput Netw 54(15):2787–2805
39. Mashal I, Alsaryrah O, Chung T-Y, Yang C-Z, Kuo W-H, Agrawal DP (2015) Choices for interaction with things on Internet and underlying issues. Ad Hoc Netw 28:68–90
40. Ray PP A survey on Internet of Things architectures. J King Saud University Comput Info Sci
41. Guth J, Breitenbücher U, Falkenthal M, Leymann F, Reinfurt L (2016) Comparison of IoT platform architectures: a field study based on a reference architecture. In: 2016 cloudification of the Internet of Things (CIoT), pp 1–6
42. Gubbi J, Buyya R, Marusic S, Palaniswami M (2013) Internet of Things (IoT): a vision, architectural elements, and future directions. Future Gener Comput Syst 29(7):1645–1660
43. Bedek MA, Firssova O, Stefanova EP, Prinsen F, Chaimala F (2015) User-driven development of an inquiry-based learning platform: qualitative formative evaluations in weSPOT. Interact Design Architecture(s) 23:122–139
44. MongoDB for GIANT Ideas [Online]. Available: https://www.mongodb.org/
45. Apache CouchDB [Online]. Available: http://couchdb.apache.org/

Trust Management Approaches in Mobile Adhoc Networks

R. Vijayan and N. Jeyanthi

Abstract Mobile Adhoc Networks are a collection of mobile nodes interconnected by wireless media links. Here nodes misconduct owing to topology variations, exposed medium, short signal range, the absence of centralized approach and limited Energy. Node misbehaviors are critical concerns for routing in MANET. Under the given scenario, the nodes in MANET might decline to collaborate by not forwarding data with others for egoistic causes. However, it mandated that the nodes must collaborate or cooperate to route the packets towards the appropriate destination. The trust relationship among the nodes in a disseminated way helps in detecting the misbehaviors/selfish nodes. To avoid the misbehaving nodes from routing three trust management approaches has been proposed. In the first approach, Trust approach for Discovering and Quarantine the Misbehaviors (TDQM), the number of dropped packets as a parameter in the neighbor observation trust evaluation. Energy utilized for forwarding all types of packets, packet integrity, and neighbor observation, trust is recommended for trust evaluation. A node with high trust elected as a certification agency to release certificates for routing packets between the desired nodes. Moreover, untrustworthy nodes trust, computed is redeemed and isolated from routing. A Second approach proposed is Context residual Energy Based Trust Management in MANET (CEBTM), Retrieved context data like residual Energy and packets dropping of the nodes considered with weights for neighbor observation trust evaluation and reputation-based trust based on fuzzy for analysis. Revocation of trust of a generated negative alarm is verified based on the probability of reliability. In the Enhancing Trust in a MANET using trust and dynamic with Energy Efficient Multipath Routing Protocol approach (ETMRP), evaluating nodes parameters intimacy, Energy, packet delivery and co-ordination as context values over time. Here trust

R. Vijayan (✉)
Department of Information Technology, School of Information Technology and Engineering, VIT University, Vellore 632014, Tamil Nadu, India
e-mail: rvijayan@vit.ac.in

N. Jeyanthi
Department of Computer Application and Creative Media, School of Information Technology and Engineering, VIT University, Vellore 632014, Tamil Nadu, India
e-mail: njeyanthi@vit.ac.in

© Springer Nature Switzerland AG 2019   69
N. Jeyanthi et al. (eds.), *Ubiquitous Computing and Computing Security of IoT*,
Studies in Big Data 47, https://doi.org/10.1007/978-3-030-01566-4_4

based path chosen by Energy efficient multipath routing. The simulation results of the proposed approaches provide healthier packet delivery ratio, reduced end-to-end delay as compared with other methods.

Keywords Trust management · Trust framework · Trust misbehaviors
Context-based trust · Trust computation in MANET
Trust and dynamic with energy efficient multipath routing protocol

1 Introduction

The concept of trust is necessary to communicate, and network protocol designers where establishing trust relationships among participating nodes is critical in enabling collaborative optimization of system metrics. Trust as "a set of relations among entities that participate in a protocol". The attackers also attract the unique features of the vulnerable medium, changing topology, limited battery power, limited physical security MANETs. In common, uncooperative nodes in MANETs may be of two types: malicious nodes and selfish nodes. A self-regarding node, on the other hand, is an economically rational node whose objective is to maximize its welfare, defined as the benefit of its actions minus the cost of its operations. A node consumes some cost for forwarding a message; a selfish node will need an incentive for doing it [1].

The term Trust originally derives from social sciences and is defined as the degree of subjective belief about the behaviors of a particular node [2]. Matt Blaze announced the word Trust Management and recognize it as a detached constituent of security services in networks and clarified that Trust management provides a unified approach to specify and interpret security policies, credentials, and relationships.

Trust is definite as a degree of belief about the behavior of other units. Creating trust relationships among participating nodes is vital that non-cooperative or destructive behavior will hurt their trust value or reputation. Motivation towards trust is from support in decision-making to improve security and robustness, Misbehavior detection, Adaptation to risk. MANETs must deliver numerous levels of security assurances to diverse applications for their successful placement usage. However, due to their wireless links and lack of central management, MANETs have far more significant security concerns than conventional networks. There is a need for the node in MANET to be trusted or not for various applications, route discovery and other. Multiple attacks exist in MANET. An active attack is attacking when a misbehaving node has to bear some energy costs. On the other hand, passive attacks are mainly due to lack of cooperation with the purpose of saving energy selfishly. Nodes that perform ongoing attacks with the intention of harming other nodes by causing network outage are measured as malicious while nodes that make passive attacks with the aim of saving battery life for their communications are deemed to be selfish.

2   Literature Review

CONFIDANT is a mechanism which supports cooperation in ad hoc networks by detecting and isolating malicious nodes using direct observations and recommendations. The model cannot prevent the dissemination of false recommendation, and negative recommendation is the only information exchanged between nodes. Besides, it uses only single trust metric evaluation based on the cooperation of nodes in packet forwarding [3]. The model omits some critical evaluation metrics such as energy, delay and social properties in evaluating nodes' trust. Propose the CORE model, which has a watchdog component complemented by a reputation system that distinguishes between three types of information such as subjective reputation using observations, positive indirect reputation by others, and functional reputation using task-specific behavior. Consequently, this can lead to decreased efficiency of the system because nodes cannot exchange bad experiences from the misbehaving ones in the network. Also, CORE cannot be resilient against a ballot-stuffing attack as it leaves ways for misbehaving nodes to collude and gain low high ratings [4].

Authors in [5], has proposed RFSTrust, a trust model based on fuzzy recommendation similarity, which is presented to measure and evaluate the trustworthiness of nodes. They use similarity theory to assess the recommendation relationships between nodes. That is, the higher the degree of similarity between the evaluating node and the recommending node, the more consistent is the evaluation between the two nodes. In this model, only one type of situation considered when a selfish node attack is present. Chen et al. [6], present a fuzzy trust model for peer-to-peer networks. The design includes two phases; recommendation trust and direct trust. The primary focus of the recommendation trust phase is the extraction of the trust link and computation of recommendation trust degree using the fuzzy decision-making method. Various sets based on the fuzzy decision method are used to obtain a fuzzy trust evaluation metric, while the focus of the direct trust phase is mainly on updating the direct trust degree using peers experience and recommendation.

Context-Aware Detection is the mechanism proposed by Paul and Westhoff [7], relates nodes accusations to a unique route discovery process and a particular period. They use for monitoring a combination of un-keyed hash verification of routing messages, and misbehavior detection by making a comparison between a cached routing packet and overheard packets, thereby detecting tampering from the path request header. This approach enables the detection of several types of attack and attacker, and also rejects ineffective route information at as early a stage as possible. Li et al. [8], have used various contextual information, such as channel status, battery status, and weather condition, are collected and then used to determine whether the misbehavior is likely a result of a malicious activity or not. Luo et al. [9], Here the combines direct trust and trust recommendation information. Trust collaborative filtering to allow nodes to represent and reason with uncertainty and imprecise information regarding other nodes trustworthiness. There is no cooperation risk measurement relying on the approximate estimation of a node's behavior instead of detailed and crisp data.

Xia et al. [10], propose Trust relationship based on weight method and fuzzy logic rule for misbehavior detection. However, no attributes that influence the surrounding environment of the node like expected transmission count on lossy wireless links, energy, packet drop. Moreover, no adaptive trust-level classification of nodes. Trust computation not done for different context under different node encountered at a time. Energy Efficient and Trust-Based Node-Disjoint Multipath Routing Protocol for WSN a node separate multipath routing protocol which is vitality effective as well as trust-based [11]. Survey of different Trust based QoS mindful Routing Protocol in MANET trusts based and QoS-aware steering convention is performed. The investigation of various trust-based and QoS-aware AODV conventions that are utilizing trusted framework and trust models shown for keeping the assaults and rowdiness from vindictive nodes in the network. The execution of trust-based routing resolution has been examined that works productively and can utilize as a part of different utilizations of MANETs for enhancing the security performance of the system. AODV convention is employed fundamentally for the course revelation and course upkeep when the connection fizzles, consequently is for the most part utilized as a part of the routing resolutions [12].

3 TDQM-Trust Scheme for Discovering and Quarantine the Misbehaviors in MANET

In this scheme, direct trust calculation of a node to another node means calculating the trust value and confidence levels from the events that are directly experienced or monitored between those two nodes. Recommendation trust or indirect trust (IDT) means the trust value taken from other nodes. This confidence value is maybe a direct trust value on that node or the indirect trust value of that node taken from other nodes. In this scheme calculation of direct and indirect confidence is also considered the energy utilization factor of the node that is calculated based on the battery consumption by a node. Packet modification is a significant security issue in MANETs, so a node checked if it tries to alter or tamper the packet contents and this packet integrity factor is considered in the trust calculation making trust computation vital in the proposed framework as shown in Fig. 1.

3.1 Energy Auditor

In every node power expended for forwarding the packets to neighbor nodes. Later on, communication starts energy consumption also leads. This expenditure of energy is more for trusted nodes because they have to receive as well as forward the packets to its neighbors. However, is a case of selfish nodes energy utilization is slightly low, they only receive data packets, they will not forward packets to neighbors. So

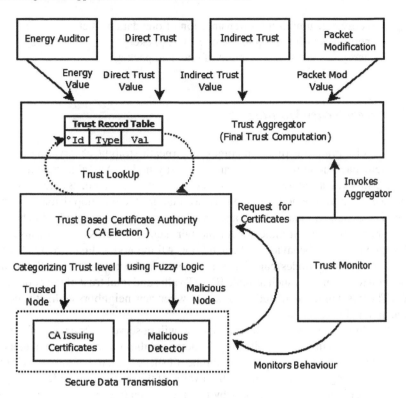

Fig. 1 Proposed TDQM trust scheme

after completion of communication trusted nodes have less energy, selfish nodes have somewhat higher energy than other nodes. This power consumption is also depending on the routing protocol.

To calculate the energy used by a node the initial node configuration is necessary. In primary node configuration initial energy, ideal power consumption, receiving power consumption, transmission energy consumption all these details should be specified.

The nodes Battery power is limited and non-renewable resource in wireless ad hoc networks. Due to limited resource availability nodes behave selfishly by saving the battery power without forwarding the packets. An energy audit monitors the energy spent at each node. The total node energy expenditure at a node due to another node in the network can calculate as follows.

$$E_{\text{Value}} = \rho_{n>0}(\rho_{X \equiv Y}E_{T_{ack}} + \rho_{X \neq Y}E_{R_{ack}}) + \rho_{m>0}(\rho_{X \equiv Y}E_{T_{Pck}} + \rho_{X \neq Y}E_{R_{PCK}}) \quad (1)$$

where

E_{Valve} = Energy consumed at node Y due to node X
E_{Valve} = Energy spent for one acknowledgment

$E_{T_{ack}}$ = Energy consumption for transmission of one data packet
$E_{R_{ack}}$ = Energy Spent for reception of one ACK packet
$E_{R_{pck}}$ = Energy spent for the reception of one data packet.

3.2 Direct Trust Monitor

Neighbor observation estimation comes underneath straight observation of neighbors. Here, each node monitors the all the nearby nodes. Any abnormal action discovered, invokes an algorithm to determine direct trust value. By neighbors of a node, we mean all the nodes in the network that are the one-hop distance from the node. The collected data audited, and any deviation from the standard behavior of a neighbor is used as an indicator for the fair degree of selfishness because this is independent of past behavior of the neighbor. All the nodes direct trust value of all the communicating nodes calculated and stored in the trust table of the resultant node with record names such as node ID, DT value and total trust of the coordinating node. The trust value calculated dynamically when new neighbors within the network range and table are updated.

In this direct trust module, the source node will observe the behavior of the target node. Here observing means collecting the details of the communication of that target node. After collecting the total details number of successful, total numbers of failures calculated. Using those details direct trust value of the destination node is computed.

This module monitors neighbors by passively listening to their communication, every node in the network monitors the behavior of every other neighbor. Identifying packet dropped, delayed packet, forwarded packet, and this module use watchdog mechanism to check whether neighbor forwards the packets or drop them. The direct trust monitor audits the collected data. DT agent does trust computation. Node X want to calculate the trust value of node Y termed as $DT_{Value}(K)$

$$K = S - F \tag{2}$$

Here S is a function of successful communication metrics, F is a function of not successful communication metrics. Evaluation DT on Y, X has to monitor the following statistics. Parameters of the role S are Control packets forwarded, Control packets received. Parameters for function F are packet delay, packet drop. K gives the direct trust value (DT_{Value}) calculated from X on Y.

3.3 Indirect Trust Monitor

The task of the secondary trust monitor is to collect or request the trust related information of target node from the neighboring nodes. Here one more issue will come in the pictorial manner, from which neighbor, have to collect this trust information.

The result for this question is a direct trust value of the neighbor. In other words, the trust request of a sink node from neighbors, the trust value of that neighbor node considered. Choosing of the donor is also done using fuzzy logic. This information called for indirect trust (IDT).

Algorithm:

Obtaining Indirect Trust on y from n

i.   Node x sends IDTREQ to node(s) n.
ii. *If* node n has direct trust value to y, then it will reply back with IDTREP.
iii. *Else If* and does not bear a direct trust value record it will discard the IDTREQ
iv. After receiving an IDTREP reply from neighbors consider the trust value of the node with maximum direct trust value by applying fuzzy logic.
v.   Integrate all the obtained IDTvalue from neighbors to calculate the Indirect trust value.

Indirect trust means getting the trust value of required nodes from its neighbors. A recommendation, trust value, the source node will broadcast the request packet regarding the recommendation trust to all its neighboring nodes. Here input of the fuzzy logic is a direct trust value of all the replied neighbors. Among all the neighbors, the node with maximum trust value chosen, and recommended trust value considered for the evaluation indirect trust value of X on node Y denoted as IDTValue.

3.4   Packet Modification Auditor

Packet modification by malicious nodes is a significant security threat. If an intermediate node modifies the message content, this amendment can be detected, and the packet discarded. PM_{value} is a positive value at the initial stage, and if there is a modification of request packets by a node, its PM_{value} will decrease. Each node generating a message includes a digital signature generated from its private key and all fields except the hop count field. RSA performs signature verification efficiently and incurs least cost compared to other asymmetric algorithms. The vast discrepancy in the energy costs of sign and verify operations in RSA results from the significant difference in the sizes of the keys employed. In our proposed scheme RSA algorithm is implemented due to its efficiency and least energy cost over others.

3.5   Trust Aggregator

The functionality of the trust aggregator is to calculate the final trust value FTvalue of the target node using a direct trust, value and recommended trust from the neighbors, Energy value and Packet modification value. When the timeout expires, Trust

Computation carried out again. Each node the trust value a timestamp assigned Trust aggregator generates a Trust Record Table (TRT) with Node id, trust type and Trust value of each node.

$$FT_{Value} = DT_{value} + IDT_{value} + PM_{value} - E_{value} \qquad (3)$$

where

FT_{value} = Final trust value
E_{value} = Energy value
DT_{value} = Direct trust value
IDT_{value} = Indirect trust value
PM_{value} = Packet modification value.

3.6 Certificate Authority

The Central Authority CA will be responsible for generating and issuing a certificate to the requesting node. For every new session, a new certificate would generate with a new timestamp TS. The node with maximum trusted value elected as certificate authorities. Once identified the source node can transmit its ID and obtain certificates from the central authority and begin transmission to the destination node. The trust value or timestamp expires the node's certificate renewed by the Certificate Authority.

The security provided by electing Trust based Certificate Authority using the highly trusted node in the range. Certificates are issued by the CA, evaluating the trust value of each node. This framework ensures only the trusted nodes to acquire certificates for data exchange isolating the malicious nodes in its range making the system trustworthy.

In this scheme, the CA Election algorithm devised, to elect CA node from the group of nodes based on the nodes trust value. Each node sends a CA request packet (CAREQ) to announce it as CA. CAREQ consists of node ID and trust value. When a node of confidence receives a CAREQ, from one of its neighbors, choosing CA. A node with the high trust value elected as a CA. A single mobile node functioning as a CA will bring the entire MANET to a halt if it moves out of the MANET and also acts as a single point of failure if it becomes compromised. Replacement CAs is being used to prevent this security bottleneck.

CA Reply packet (CAREP) sent back to the requestor node. It consists of node ID and corresponding nodes trust value. If the requestor node elected as CA node, it maintains the trust value received from the other neighboring nodes and highest among them will be replacement CA. The Timeout value set to the CA node. When the timer value, perishes or CA node fails, CA election algorithm will be invoked again.

Table 1 Fuzzy discrimination

Fuzzy levels	Trust values	Semantics
1. Very high	0.8–1	Highly trustworthy
2. High	0.6–0.8	Medium trustworthy
3. Medium	0.4–0.6	Trustworthy
4. Low	0.2–0.4	Untrustworthy
5. Very low	0–0.2	High untrustworthy

Table 2 Fuzzy rules

1. IF trust value is VERY HIGH	THEN node is TRUSTED
2. IF trust value is HIGH	THEN node is TRUSTED
3. IF trust value is MEDIUM	THEN node is TRUSTED
4. IF trust value is LOW	THEN node is UNTRUSTED
5. IF trust value is VERY LOW	THEN node is UNTRUSTED

3.7 Trust Categorization Using Fuzzy

Fuzzy logic based algorithm for trust has devised and applied to the calculated trust value of the nodes. Trust values are computed based on E_{value}, T_{value}, PIC_{value} produces FT_{value}. These values treated as fuzzy input variables and the Fuzzy logic based algorithm marks the nodes as either trusted or malicious. A two-way Fuzzy based analyzer has been designed based on trust values, either trusted or noticeable as malicious less than the threshold and its isolated from the network. Table 1 categorizes the trust levels based on the fuzzy theory of computation [13].

Fuzzy inference rules for categorizing the nodes based on trust levels as shown in Table 2.

A node requests the CA for certificates to perform a data exchange. Fuzzy Based Analyzer verifies the trust value of the requesting node and performs a lookup in the fuzzy table for the fuzzy trust value. Fuzzy Based Analyzer runs the algorithm to determine the node as trusted or malicious. The CA node finds a requesting node as malicious an alarm generated to the intimate malicious node to all the trusted nodes in its range. The network is secured by detecting and isolating the malicious node and prevents them from performing any activity in the range.

3.8 CA Issuing Certificates

After the source and destination nodes obtain certificates from CA, it is eligible for packet transmission. Source node uses the public key to hash the packet and forwards it to the destination. Only the target node can verify the packet using its private key. Hash algorithms are least complex for cryptographic algorithms and should incur least energy cost. In this scheme, MD4 is used to hash the packet. Once the timestamp value of the certificate expires, the node has to request CA node for the renewal of licenses.

Certificates issued by the CA node with a timeout value, and once the timeout value of the trusted node expires, it has to request the CA node for the renewal of certificates to transmit data packets.

Certificate Exchange Algorithm:

i.   Generate Shared Key SKac.
ii. Source node request CA E[CREQ(SID,DID,FTValue)SKac]
iii. CA node decrypts CREQ looks for SID in ID repository.
iv. IF(SID==ID)THEN
v.   CA node verifies for SID and checks for DID in its range.
vi. Generate PUx,PRx, PUy,PRy, SKbc,

    a.   CERT X = SID,PRx,PUy,FTvalue,TS.
    b.   CERT T = DID,PRx,PUy,FTvalue,TS.

vii. CA sends CREP as E[(CERT X)SKac] to source node A.
viii. CA sends E[(CERT Y)SKbc] to destination node Y.
ix.   ELSE DISPLAY("Transmission not granted").

where
PUx, PUy = Public Key of node A and B.
PRx, PRy = Private Key of node A and B.
SKac = Shared Key of Source and CA.
SKbc = Shared Key of Destination and CA.
SID, DID = Source and Destination ID.

3.9 Malicious Node Detection

Nodes with the fuzzy values as low, very low marked as malicious. Fuzzy Based Analyzer invokes the Fuzzy logic based algorithm to detect the malicious nodes. CA node denies the certificate to the malicious nodes, preventing them from participating in the network activities. An alarm is generated by the CA node to indicate the node's malicious behavior to other trusted nodes in its range, thus isolating the less trusted nodes and building a secure system. No suspicious and misbehaving nodes can cause vulnerabilities and threats to the proposed scheme. The proposed scheme is made secure by incorporating trust levels and Fuzzy Based Analyzer for Certificate Authority. Fuzzy Based Analyzer performs the defined steps, and if the requestor node is TRUSTED, then CA node generates the certificates and sends it to the requestor node. Nodes with the fuzzy values as VERY HIGH, HIGH, and MEDIUM fall in the TRUSTED category. Now with the help of the acquired certificate, the TRUSTED node can exchange the data packets. Certificates issued by the CA node with a timeout value and once the timeout value of the TRUSTED node expires it has to request the CA node for the renewal of certificates to transmit data packets.

4 Simulation Results and Discussion

Qualnet5.0 network simulator is used to simulate a wireless network with AODV protocol. A network model has created as shown in the figure with the help of Qualnet Simulator. Six nodes have used in this scenario where Node ID 1, Node ID 2, Node ID 3, Node ID 4, Node ID 5, Node ID 6 connected to wireless network framing a MANET. Node ID 5 is assumed as central authority node and receives the message from both the source and the destination node. An AODV routing protocol has used in this scenario wherein every node exchanges routing information with the other nodes in the system. Random waypoint mobility has set with all the nodes, and the behavior of the nodes with mobility is displayed.

The Simulation parameters for six nodes model as shown in Table 3. The packet sent for this model as shown in Fig. 2 and packet received in Fig. 3. The packet dropped nodes for this model as shown in Fig. 4.

The Simulation parameters for twelve node model as shown in Table 4.

Secondly, network model has created as shown in Fig. 5 with the help of Qualnet Simulator twelve nodes have used in this scenario. Here Node ID 1–12, are connected to wireless network framing a MANET. Node ID 12 is assumed as central authority node and receives the message from both the source and the destination node. An AODV routing protocol has used in this scenario wherein every node exchanges routing information with the other nodes in the system. Nodes without mobility in this scenario and probability of detection of malicious nodes with the proposed

Table 3 Simulation Parameters-Scenario-I

Parameter	Values
MAC	IEEE802.11
Routing protocol	AODV
Initial Energy	100
Reception Power	1.040 W
Transaction Power	1.6787 W
Idle Power	0.6699 W
Simulation time	1 min
No. of nodes	6
Mobility Model	Random waypoint
Traffic type	CBR
Payload size	512 bytes

Table 4 Simulation Parameters-Scenario-II

Parameter	Values
MAC	IEEE802.11
Routing protocol	AODV
Simulation time	1 min
Initial Energy	100
Reception Power	1.049 W
Transaction Power	1.6787 W
Idle Power	0.6699 W
No of nodes	12
Mobility Model	Random Waypoint
Traffic type	CBR
Payload size	512 bytes

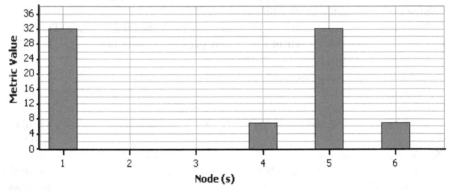

Fig. 2 Number of data packets sent for six node model

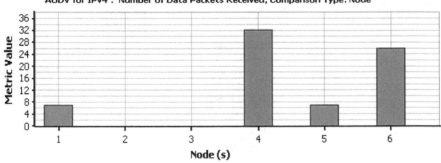

Fig. 3 Number of data packets received for six node model

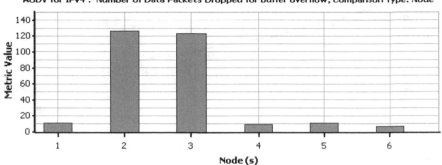

Fig. 4 Number of data packets dropped for six node model

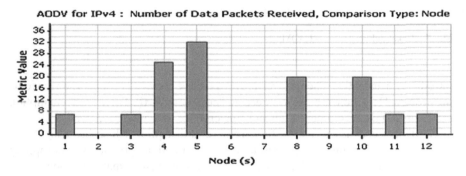

Fig. 5 Number of data packets received for twelve node model

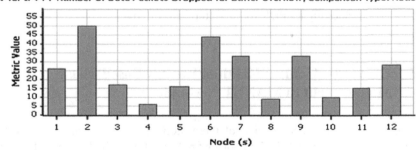

Fig. 6 Number of data packets dropped for twelve node model

framework depicted. A random waypoint mobility with all the nodes and the behavior of the nodes with mobility is displayed as shown in Fig. 6.

The effect of implementation of our approach, Nodes two, three in six model and Node 2, 6, 7, 11 in 12 Node model marked as malicious and isolated.

CONFIDANT determines the value of trust using direct and indirect monitoring. By these observations, it detects the malicious nodes. CONFIDANT additionally

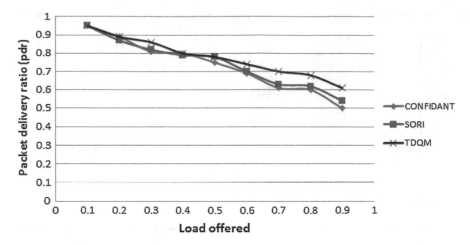

Fig. 7 Packet delivery ratio—TDQM versus other trust approaches

introduces an incentive scheme to reward the positive nodes that cooperation the routing process. Here there is more overhead and complexity exit. There is no redemption mechanism, fake recommendation and convergence of the iteration persist.

SORI that encourages packet forwarding and discipline selfish behavior. Interaction in a one-hop range in promiscuous mode to find the probability of dropping packets based on a threshold. However, does not prevent a malicious node from selectively forwarding packets or the other malicious behavior.

In Fig. 7 compares our proposed TDQM: Trust scheme for Discovering and Quarantine the Misbehaviors with CONFIDANT and SORI. It observed that our approach TDQM provides higher PDR for more the load offered.

5 CEBTM—Context Residual Energy Based Trust Management in MANET Network

The proposed approach consists of three modules as shows in Fig. 8, with each playing an important role in deriving the final trust for a neighbor node. The modules are Context Data Manager, Trust Manager, and Trust Monitor. The Context Data Manager that takes care of two major operations as sending context data onto neighbor nodes and getting context data onto neighbor nodes and computing the weight for each node at a particular context data. The Trust Manager computes final trust value, rating on the threshold and triggering a negative alarm to Trust monitor. The Trust Monitor verifies the alarm for validity intimates to all neighbor nodes.

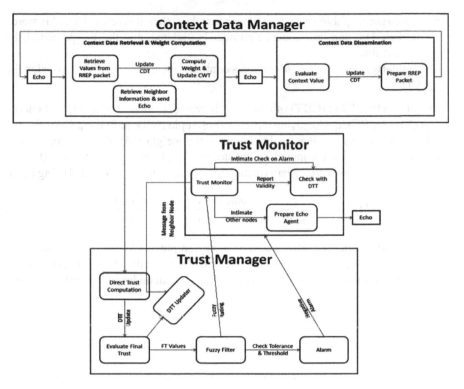

Fig. 8 CEBTM model

5.1 Context Data Manager

The proposed system dynamically computes trust by taking the context data like Residual Battery energy, Packet Drop Strategy of a node and sending these details to the neighbor nodes of the network. These details received via a handshaking agent named Echo which as shown in the pseudo code below:

Pseudo code for context data evaluation and distribution:

 i. Every node identifies its neighboring nodes.
 ii. Each node finds its context data, i.e., Residual battery energy ($C_{re,i}$) and Packet drop strategy ($C_{pd,i}$) and store it in the Context Data Table (CDT).
 iii. The sequence numbers (RESeq & PDSeq) with zero and the fraction of a lifetime updated in the F_t.
 iv. Attaching Residual energy value of the source to the Echo agent and send it to all the neighbor's destination nodes with an initial sequence number as zero.
 v. Attaching Packet Drop Strategy value of the source to the Echo agent and send it to all the neighbor destination nodes with an initial sequence number as zero.

  vi. If a target neighbor sends back its context data, then it is retrieved and is stored in the CDT, and the sequence number is updated, and the F_t updated at the current time.

 vii. Similarly, the entire neighborhood's context data are retrieved and stored in the CDT.

Context Data Table (CDT) of a node 'i' is represented as a node residual energy tuple (NID_{ires}, REseq, $C_{res,i,Ft}$) and packet drop tuple (NID_{ipd}, PDSeq, $C_{pd,i}$, F_t).

The context data are then used to compute the weight [14] of each node about a particular context. The computed weight stored in the Context Weight Table (CWT). The CWT gets updated after every fraction time. The weight is computed using Eq. 4.

$$Weight, \ W_{i,c} = \frac{Context \ data \ 'c' \ of \ node \ 'i'}{Summation \ of \ values \ of \ all \ nodes \ of \ context \ 'c'} \qquad (4)$$

where 'i' is the node for which weight is to be computed, c' is the context data.

This weight of a node for a particular context is the ratio of the context value of the node to the summation of all the context values of all nodes to particular contexts.

5.2 Trust Manager

The weight of each node in a particular context is then used to compute the trust in each node in the specific context.

Context Weight Table (CWT) of a node 'i' is represented as a tuple (NID_i, $W_{i,res}$, $W_{i,pd}$),

where, NID_i—Node ID, $W_{i,res}$—Weight for the context Residual Energy, $W_{i,pd}$—Weight for the context Packet Drop Strategy

$$\text{Direct Trust of node 'j' on 'i'} \quad DT_{i,j}^c = \sum_{c=1}^{n} W_{i,c} \times F_t \qquad (5)$$

$$\text{Fraction Time}, \quad F_t = t_c^{st} \Big/ \sum_{c=1}^{n} t_c^s \qquad (6)$$

$$t^{st} = \text{current time} - \text{FLT for a particular Node 'j'} \qquad (7)$$

where i, j is nodes in the network, 'c' is the context data Residual Energy and Packet Drop Strategy, F_t is the fraction time. The Direct trust thus computed is then shared with the neighbor nodes and t_c^s is the summation of all t^{st} for a context data 'c'. For every 180 s, the context data are shared with the neighbors, and the CWT gets updated. The direct trust [15] value that is computed using Eq. 5 by each node about the context data that it represents is computed. The direct trust value computed is stored in the Direct Trust Table (DTT). The DTT is maintained by each node in the network regarding the neighbor nodes.

Direct trust table (DTT) of a node 'i' is represented as a tuple (NID$_i$, DT$_{i,j}^{res}$, DT$_{i,j}^{pd}$, FSTV$_i$),

where NID$_i$—Node ID, DT$_{i,j}^{res}$—Direct Trust for Residual Energy, DT$_{i,j}^{pd}$—Direct Trust for Packet Drop Strategy, FSTV$_i$—Final Source Trust Value.

The average of the direct trust computed for each context data are calculated, and the resulting value is the final source trust value of a node calculated by it. This final source trust value stored in the DTT. The final source trust values of each node along with its node IDs are then shared with the neighbor nodes via the Echo agent and stored in the Final Node Trust Table (FNTT) of the node.

Final node trust table (FNTT) of a node 'i' is represented as a tuple (SNID$_i$, FSTV$_j$, FNTV$_{j,i}$), Where SNID$_i$—Source Node ID, FSTV$_i$—Final Source Trust Value of a node 'j', FNTV$_{j,i}$—Final Node Trust Value.

In FNTT, the Source Node ID is the Node ID from which the Final Trust Value sent, and Final Source Trust Value of a node 'j' is the node to which the Source Node has computed the Final Source Trust Value. Finally, the average of the Final Node Trust Values of each neighbor node and its own Final Source Trust Node is computed using Eq. 8 and is stored in a Final Trust Table (FTT).

$$Final\ Trust\ Value = \frac{FSTV_i + FNTV_{j,i}}{n} \tag{8}$$

Final trust table (FTT) of a node 'i' is represented as a tuple (NID$_i$, FTV$_{i,}$).

where, NID$_i$—Node ID, FTV$_i$—Final Trust value of a node 'i', The FTT is then sent to the Fuzzy Filter to check whether the computed final trust can be trusted or not.

5.3 Fuzzy Analysis

The fuzzy filter analyzes the Final Trust calculated by the node and done in a series of steps shown below. The Final Trust Value computed is checked with the Reliability Table.

Pseudo code for final trust verification:

i.   The Context data and the Final Trust Value then verified with the Reliability table, and the reliability value is derived.
ii. If the Reliability value is 1.0, then the node is most trusted of.
iii. Else if it is between 1.0 and 0.5, then the node can be trusted.
iv. Else if the Reliability value is 0.4, then the node can be trusted with some exceptions.
v.   Else if the Reliability value is less than 0.4, then the node is not trusted the untrustworthy node is sent for revocation to the Trust Monitor of other neighbor nodes via Echo agent.

The reliability of node j at time t, denoted by $R_j(t)$, is the probability that node j meets the required trust level for mission execution over time [0, t], calculated as follows [16]:

$$R_j(t) = \{0\} \quad if \ T_j(t') = 0 \ \ for\ any\ t' \leq t \tag{9}$$

$$R_j(t) = \{E[T_j(t')]\} \quad for \ \ t' \leq t \ \ otherwise \tag{10}$$

In Eqs. 9 and 10 where t is the current time point and t' $\leq$ t is a past time point as shown in Table 5.

5.4 Trust Monitor

The untrusted node that reports the untrusted is verified with the trust value that it has computed on its own. For example, if node 'i' reports to its neighbor that node 'j' is non-trusted to neighbor node 'k', then node 'k' analyses the trust which it has computed on node 'i' moreover, if the reliability value is above 0.7, then node 'j' is declared non-trusted, and the CDT updated as 'null.'

5.5 Echo Agent

The Echo agent is a two-way handshaking agent that exchange the context data or information between a source node and a destination neighbor node. The Echo agent is one hop away and is modified by each destination node to send it back to the node of origin with the relevant data.

The Echo agent used for data exchange in sending context data to neighbor nodes, requesting context data from neighbor nodes, sending final trust of the source node to neighbor nodes and intimation neighbor node about the untrustworthy node.

Table 5 Reliability Table

Trust value	Packet drop strategy	Remaining energy	Reliability	Reliability value
9–10	80–100	80–100	Very very High	1.0
7–8	80–100	80–100	Very High	0.8
5–6	50–79	50–79	High	0.6
3–4	50–79	50–79	Medium	0.4
1–2	50–100	50–100	Low	0.2
0–1	00–40	00–40	Very low	0.0

5.6 *Simulation Results and Discussion*

The proposed framework is simulated using Network Simulator 2 (NS2) by editing the existing AODV routing protocol. The Echo agent after being developed patched with the NS2.35 and the simulation handled with mobile node scenarios. The scenarios are developed under NS2.35 and are simulated under Linux environment Ubuntu 10.05. The resulting final trust is plotted using gnu plot to show the level of trustworthy and untrustworthy nodes in the simulated scenario. Table 6 shows the scenario I, where the 15 mobile nodes considered. The scenario uses energy model to analyze the amount of residual energy present in the node. All the energy parameters specified in the table.

Table 7 shows the scenario II, where the 75 mobile nodes considered. The scenario uses Energy model and other mobility models to analyze the amount of residual energy present in the node and the mobility of the particular nodes respectively. All the energy parameters specified in the table.

Table 6 Simulation Parameters for Scenario I

Parameter	Value
MAC	IEEE 802.11
Routing protocol	AODV
Models included	Energy model
Initial energy	100
Simulation area	300×300
Reception power	35.28e−3
Transaction power	31.32e−3
Idle power	712e−6
Nodes	15 with mobility

Table 7 Simulation parameters for Scenario II

Parameter	Value
MAC	IEEE 802.11
Antenna	Omni antenna
Models included	Energy model, random way point with wrapping, random way point with steady state
Initial energy	100
Simulation area	1700×1700
Reception power	1.049 W
Transaction power	1.6787 W
Idle power	0.6699 W
Nodes	75 with mobility

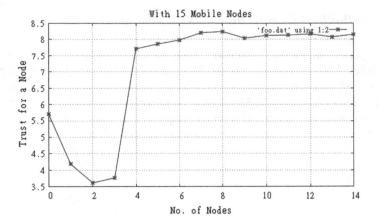

Fig. 9 Trust for scenario

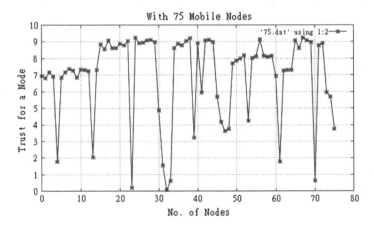

Fig. 10 Trusts for scenario II

Figure 9 shows the resultant final trust for each node for scenario I. Figure 3 shows the resultant final trust for each node for scenario II. Figures 9 and 10 shows the x-axis the Number of nodes considered for the scenario and y-axis showing the Final Trust for each node. The trust values will then compared with the reliability table, and the CDT maintained by each node to find the reliability value of each neighbor node. By analyzing the above two figures, the progressive trust values of the nodes were computed precisely for more number nodes in the MANET by considering the context parameter of residual energy and packet drop strategy.

Figure 11 compares our proposed CEBTM: Context Residual Energy Based Trust Management in MANET with CONFIDANT, CORE, and TAODV. It observed that our proposed approach provides higher PDR for more the load offered.

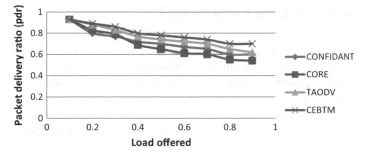

Fig. 11 Packet delivery ratio versus load offered

6 ETMRP—Enhancing Trust in MANET Using Trust and Dynamic with Energy Efficient Multipath Routing Protocol

This segment gives a detailed description of the proposed Trust based Routing convention. The trust-based random energy efficient multipath routing is proposed to be a secure and random routing directing strategy that gives security and fruitful load appropriation to every one of the nodes in the network.

Trust is standard for a node that is a measured representation of satisfaction estimations on various hubs (node) in the framework and is made in light of affiliation experiences with them at a particular time. Trust assessed on different measurements and unique ways. A few plans utilize consistent or discrete qualities to quantify the level of trust. For Example, trust is depicted by a consistent esteem in [0, 1] or measured as the discrete esteem in [−1, 1]. Edge-based methodologies are likewise used to quantify the trust. Trust measurements, for example, fluffy based, likelihood-based, closeness, portability, setting based elements like energy, signal strength, hop distance and other. The trust has been estimated and handling by using the three different sections, like, "experience," "Proposal" and "learning." The "experience" a portion of trust for each center point is straightforwardly measured by their speedy neighbors and kept redesigned at ordinary between times in the trust table.

The present trust table is spread with every other center point as "suggestion" a bit of the trust. At a regular break, then as of now evaluated a trust joined in the present "Learning" section of total trust. In a matter of seconds either these three portions autonomously or a mix of them used as a piece of processing the trust. Each created trust in the network has several properties such as, trust choice system ought not to work under the presumption that all hubs are helpful for MANETs, trust must resolve in an intensely flexible way without over the top calculation and correspondence stack. Trust is not static; it is alterable, subjective, not definitely transitive. Based on the trust properties, the trust has distributed in the network which explained as follows.

6.1 Distributed Trust Evaluations

Distributed trust assessment can be named: neighbor detecting (coordinate trust), proposals based trust (roundabout trust), and a half and half technique [17]. In conveyed specially appointed systems, trust levels conceived from the investigation of gathering information from perceptions for particular activities. It could log that a specific node advances a few packages as typical, and after that drops different bundles. It could get this through direct neighbor detecting and figure trust from direct experience. Confidence among prompt neighboring hubs is known as immediate trust and required for situations where a trust relationship framed between two hubs without past communications. Then the connection between the node helps to determine the trust level assessment from the hub and about Node b, $T_a(b)$ (as its very own weighted total trust (monitor) and the proposals of neighbors. The essential condition is

$$T_a(b) = (1 - \alpha)E_a(b) + \alpha R_a(b) \tag{11}$$

where the variable $E_a(b)$, that ranges from [0,1], says to the trust hub a has on node b construct just with respect to its own particular perceptions and $R_a(b)$, that reaches from [0,1], is the total estimated of the proposals from every single other neighbor, The variable α, that reaches from [0, 1], is a parameter in our model that permits nodes to choice the most pertinent element. The estimation of (b) is given by

$$E_a(b) = \beta E_T(b) + (1 - \beta)T_a(b) \tag{12}$$

where (b) expresses to the trust esteem got by the judgment of the activities of a neighbor performed by the Classifier segment, and the variable $E_T(b)$ gives the last trust level esteem put away in the Trust Table. The variable β, extents from [0, 1], that permits distinctive weights for the elements of the condition, selecting which component is the more pertinent at a given moment. Equations 11 and 12 depict how the Trust Calculator consolidates the data from the Experience Calculator $((b))$, the Recommendation Calculator $(R_a(b))$, and the Trust Table $(T_a(b))$ to determine a trust level. The formed trust has been managed by using the following phases which avoids the intermediate attacks present in the network.

6.2 Trust Management

The developed trust has been organized and administered by applying the Trust computations, Trust aggregation, Trust propagation and Trust prediction stages which explained as follows.

6.2.1 Trust Aggregation

Like most trust aggregation protocols for MANETs, consider both direct trust and roundabout trust. That is, the hub assesses hub j at the time by direct perceptions and circuitous suggestions. Coordinate perceptions are immediate confirmations gathered by hub toward hub j over the time interim $[t - d\delta t,]$, when hub i and hub j are 1-bounce neighbors at time 't' (Parvathi Raj and Krishnan [18]). Here Δt is the trust overhaul interim and dis an outline parameter determining the extent to which late cooperation encounters would add to intimacy. Go back similarly as t = 0, that is, d = $t/\Delta t$, if all connection encounters are considered similarly essential. Roundabout recommendations are aberrant confirmations given to hub I by a subset of 1-bounce neighbors chose in light of two systems against slandering attacks: edge based separating by which just reliable recommenders with trust higher than a smaller than normal mum trust edge are qualified as recommenders, and (b) significance based trust by which just recommenders with high trust in trust segment X are qualified as recommenders to give suggestions about a trustee's trust part X.

The node to figure its push toward hub j, $T_{i,j}(t)$ where X is a trust segment by:

$$T_i(t) = \beta_1 T_{i,j} \, direct, \, X(t) + \beta_2 T_{i,j} \, in \, direct, \, X(t) \tag{13}$$

Now Eq. 13, β_1 is a parameter to measure node i's data toward hub 'j' at time t, i.e., "coordinate observations" or "self-data" and β_2 is a parameter to weigh backhanded data from recommenders, i.e., "data from others," with $\beta_1 + \beta_2 = 1$.

6.2.2 Trust Propagation

Trust estimation on a specific node by whatever other node brings about a cost of assets. These assets, particularly in MANETs, can be constrained. A specific end goal to lessen assets spent on reconsideration of trust by different hubs decreased if the figured trust gets propagated in the system Trust engendering can be of multi-bounce. Trust proliferation depends on the transitive property of the trust. The center variable considered for trust propagation is collaboration in the system in transporting the visitor data.

6.2.3 Trust Prediction

Trust prediction is a technique for anticipating conceivably deep trust between nodes utilizing the present and past conduct of nodes. Furthermore, the suggestions got from different nodes. A trust forecast framework, a necessity ought to be fulfilled, for example, if an exchange is unsuccessful, however, the framework predicts that it would be efficient results in a false positive where a fruitful exchange that is predicted to be unsafe by the framework is a false negative.

6.2.4 Trust Formation

Characterize trust parameters utilized in the trust development convention design. The framework can find and apply the best trust arrangement parameters to boost application execution, given the operational profile as information. The subjective trust estimation of (node) hub j as assessed by the hub at time t, meant as $T_{i,j}(t)$ along these lines is processed by hub a weighted normal of closeness, wellbeing, vitality, and coop-helpfulness trust parts. The evaluation is done occasionally in each Δt interim.

Particularly node, it will process $T_{i,j}(t)$ by:

$$T_{i,j}(t) = \sum YT_{i,j}(t)X \tag{14}$$

where $T_{i,j}(t)$ is the trust, conviction of node 'i' toward node join trust segment $Y =$ intimacy, strength, vitality or helpfulness and X is the weight connected with Y. Underneath we utilize the documentation $w_1{:}w_2{:}w_3{:}w_4$ for *intimacy:ealtiness:energy:cooperativeness* for notational comfort. For a trust-based application, the best setting of $w_1{:}w_2{:}w_3{:}w_4$ exists to amplify the application execution. For a non-part, say, node j, the trust level $T_{i,j}$(t) is the same as its trust level at the last trust assessment instant $t - \delta t$ reduced by time rot.

An interesting metric is the general average "subjective" trust level of hub j, meant by (t), (s assessed by every dynamic hub. When we acquire $T_{i,j}(t)$ from Eq. 14, $T_{jsub}(t)$ can be figured by:

$$T_{jsub}(t) = \sum T_{i,j}(t) \tag{15}$$

Consider $T_{jsub}(t)$ with the "objective" trust of center point j, connoted by $T_{jobj}(t)$, figured in light of bona fide, overall information to see how much deviation subjective trust appraisal is from target trust evaluation. Specifically, let (t) show the "objective" trust of center join trust portion Y at time 't', which can secure by a numerical model. Then, taking after Eq. 14, (t) is figured by:

$$T_{jobj}(t) = \sum YT_{obj}(t)X \tag{16}$$

Then the exact algorithm for the trust creation is explained as follows [11].

Algorithm
Step 1: Multipath routes from every hub are made out, and routing table for every node created.
Step 2: Calculate the trust among nodes. For this reason, first, calculate no. of fruitful packages exchanged between each match of nodes of the considerable number of ways in a day and age (Δt) indicated by succ. pkt(Δt). At that point, a figure no. of unsuccessful packets exchanged between each combination of nodes of the significant number of ways in a day and age (Δt) signified by unsucc. pkt(Δt). Currently, T_{ij}, the trust esteem between two nodes i and j for time Δt.

$$T_{ij} = \frac{10 \times succ\,pkt(\Delta t)}{succ\,pkt(\Delta t) + unsucc\,pkt(\Delta t)} \times \frac{1}{\sqrt{unsucc\,pkt(\Delta t)}} \tag{17}$$

The trust value of all the paths calculated as,

$$T_{pmi} = \sum_{i=1}^{k} T_{i,i+1} \tag{18}$$

The average trust value of each path can found as,

$$T_{pmi} = \left(T_{pmi}/k\right) \tag{19}$$

where k = total no. of nodes in a path

Step 3: The best path is then chosen by the attribute weighing method as,

$$P_w = (\alpha \times (\text{Lowest R.E. value})) + \left((\alpha - 1) \times T_{pmi}\right) \tag{20}$$

α = Constant

Step 4: This path taken by a source node for data transmission.

The MD5 hash work H is utilized to make message process H(M) at the hub of origin. The hub of origin produces the digital signature, $d_{sign} = (H(M))d \mod n$ by encoding the message process H(M) with its private key d where, $n = p * q$, p and q are random prime numbers with $p \neq q$. The source node advances de_{sign} with information M, (d_{sign}, M) to its neighboring node through the way it takes to achieve sink.

A neighboring node on the gathering of (d_{sign}, M) and the way in the information package, confirm the digital signature by looking at a decoded estimation of design mod n with message process H(M). The outline mod n is unscrambled utilizing the key (e, n) utilizing sender's open key.

Outline mod $n = ((H(M))^d \mod n)^e \mod n = (H(M))^{ed} \mod n$ By applying Little Fermat's and Chinese Remainder Theorem to Eq. 20, and verified that

$$d_{sign}^e \mod n = H(M) \tag{21}$$

Step 5: This procedure reworked in each hop of the node-disjoint way amongst source and goal.

After creating the trust between the nodes present in the system, the information should transmit in the network for avoiding the security and energy issues. Then the transmission is done by applying the Energy Efficient Multipath Routing Protocol.

6.3   Energy Efficient Multipath Routing

The multipath routing initially used in wired networks for its reliability and its ability to balance traffic load over the network. In recent years, such technique is extended to wireless ad hoc and sensor networks with objectives to achieve better energy efficiency and network robustness in case of node failures. The multipath routing protocol is utilized to locate various disjoint ways between a couple of sink and source nodes. It has three stages, the initialization phase, the paths search phase, and the data transmission and paths maintenance phase.

6.4   Initialization Phase

The hello message is one of the control messages exchanged between nodes in the initialization phase. Identifying the neighbors and sink nodes with their forward node ID, energy and hop count using hello message. Then hello message is first exchanged between sink and source nodes, each node has the sink table and the neighboring node table updated. Each node then broadcasts a connectivity message to its immediate neighbors to specify the number of sinks that the sender is aware. This message additionally incorporates the field sink numbers to determine the number of the sink that known by the sender. The following areas give the sink IDs and the hod count to each of them altogether. The receiving node updates the corresponding entry in its neighboring node table.

6.5   Paths Search Phase

This phase initiated when a set of nodes detect the stimulus, and the selected source node begins to send the aggregated data to the sink node. Explore multiple disjoint paths, the source node unicasts one request message to every neighboring node with a distinct route ID. Each node receives the request message, updates its routing and neighbor table, and the intermediate node has to select one of its neighbors to forward this request based on that node is not selected for another path and link cost to the neighbor node has to be the lowest among all the available neighbors. The neighbor tables are updated at every hop to avoid using this path in the future path search. After the request message reached the sink node, it updates its table of sources and routing table with forwarding node ID, route ID, forward node energy and path cost. The sink node sends the assign messages to the source via each of the selected multipath. This message includes the data transmission rate assigned to each path.

6.6 Data Transmission and Paths Maintenance Phase

After multiple paths discovered, the source node begins to transmit data packets with the assigned rates on each path. Path maintenance by sink node updates the path cost in its routing table for each packet received. The sink node monitors the multiple paths for change in data rates and path cost for re-distribute the data rates to optimize the usage of network resources when path cost changes with a threshold. It then adjusts the traffic flows and notifies the source node with the assign messages if path failure occurs and hop count falls below two and sends a reset message to the source to start Path search phase.

7   Simulation Results and Discussion

This area presents an analysis of results among the EEMRP, RTSR, and ADOV the proposed. Nature utilized for recreation is NS2. The simulation results including an end to end delay, packet delivery, and packet accuracy introduced used to break down the proposed calculation with RTSR and AODV.

Measurements used for Simulation most utilized parameters as a part of MANETs are an end to end delay, for various numbers of source hubs in the system. The time taken to send a bundle over the system is known as an end to end delay. The proportion between the bundles sent by the source node and the packages got by goal node effect is known as packet delivery ratio as shown in Fig. 12 and Table 8.

The End-to-End Delay (ETED) defined by the time taken for a packet to reach the node as shown in Fig. 13 and Fig. 14.

Packet delivery ratio (PDR) is the percentage of node successfully receive the packet at much node speed and node density.

Figure 15 displays the PDR among AOTMDV, LTB-AOMDV, and ETMRP

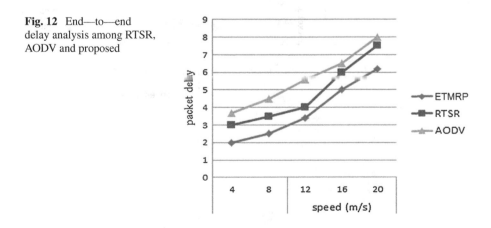

Fig. 12 End—to—end delay analysis among RTSR, AODV and proposed

Table 8 Simulation parameters for EDMRP

Parameter	Value
MAC	IEEE 802.11
Antenna	Omni antenna
Models included	Energy model, random way point with wrapping, random way point with steady state
Initial energy	100
Simulation area	1700 × 1700
Reception power	1.049 W
Transaction power	1.6787 W
Idle power	0.6699 W
Nodes	75 with mobility

Fig. 13 Packet delivery ratio among RTSR, AODV, and ETMRP

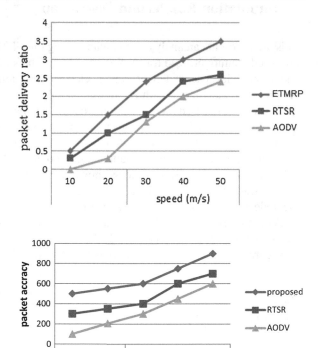

Fig. 14 Packet accuracy comparison among RTSR, AODV, and ETMRP

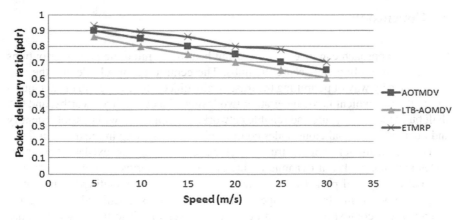

Fig. 15 Packet delivery ratio among AOTMDV, LTB-AOMDV, and ETMRP

8 Comparison of the TDQM, CEBTM and ETMRP

FEATURES	TDQM	CEBTM	ETMRP
Energy considered	Yes, Direct Trust	Yes, Direct Trust	Yes, Direct Trust
Packet dropping considered	Yes, Direct Trust	Yes, Direct Trust	Yes, Direct Trust
Direct/observation and Reputation trust	Aggregating trust	Aggregating context trust	Aggregation last trust and recommendation at a time
Context parameters for trust	No	Packet dropped and residual energy at an instant of time	Intimacy, unselfishness, energy, and cooperativeness
Weighted computation	Yes for DT	Yes for context data and trust	Yes for DT and recommendation trust
Dynamic changes in Trust value	Yes	Yes	Yes
Misbehaving node detection	Yes and isolate from routing.	Yes and isolate from routing.	Yes and isolate from routing.
Trust revocation	Yes	Yes	Yes
Timestamp for computed trust value	Yes	Yes	Yes
Trust-based Multipath routing	No	No	Yes for effective packet transmission
Simulation	QualNet and NS2	NS2	NS2

9 Conclusion

In the first approach can give a sturdy using bearing in mind the particular options such as QoS, mobile and secure the hubs. The central hub would defend the info interchange by way of permitting the trustworthy objects to take part in the net, quarantine the malevolent hubs. Integrating two levels of security, Trustworthy, and CA shall build the safe, secure and reliable network. The approach will possibly identify and quarantine the malicious nodes commencing participating in the transmission.

The second model computes trust using the ever-changing context data of neighbor nodes and thus makes it dynamic to any changing scenarios. The context data as residual energy and packet drop strategy considered for computing the trust of a neighbor node using the Echo agent. The direct trust is calculated by using the weight of the context data of each neighbor and sending it to the neighbor nodes through Echo agent further gives the idea of considering of trust of the roundabout. The Final trust evaluated by aggregation of the direct trust by itself, and the trust from the neighbor nodes led into the fuzzy filter to the evaluation of the computed trust. The Reliability table helps to identify the untrustworthy nodes of the system. The neighbor nodes are acknowledged with the untrustworthy node for revocation and making the neighbor nodes aware of its untrustworthy.

The third approach is to supply MANET protocol creators with various views taking the concept of trust, an understanding of the possessions that thought-about in creating a parameter of trust. The established trust has used for a multipath routing protocol. At the time of information transmission, the Energy of the node is maintained also manage the information with secure manner for multiple paths. Then the efficiency of this ETMRP is examined using the delay, packet delivery ratio, and packet accuracy.

References

1. Sen J (2010) A distributed trust management framework for detecting malicious packet dropping nodes in A mobile ad hoc network. Int J Netw Secur Appl 2(4):92–104
2. Chen R, Guo J, Bao F, Cho JH (2014) Trust management in mobile ad hoc networks for bias minimization and application performance maximization. Ad Hoc Netw 19:59–74
3. Buchegger S, Boudec JYL (2002) Performance analysis of the CONFIDANT protocol. In: Proceedings of the 3rd ACM international symposium. Mobile Ad Hoc Network. Computing, pp 226–236
4. Li W, Joshi A, Finin T (2011) SMART: an SVM-based misbehavior detection and trust management framework for mobile ad hoc networks. In: IEEE UMBC tech report, pp 1–12
5. Luo J, Liu X, Fan M (2009) A trust model based on a fuzzy recommendation for mobile ad-hoc networks. Comput Netw 53(14):2396–2407
6. Chen H, Ye Z (2008) Research of P2P trust based on fuzzy decision-making. In: International conference on in computer supported cooperative work in design, pp 793–796
7. Paul K, Westhoff D (2002) Context aware detection of selfish nodes in DSR based ad-hoc networks. IEEE Proc Veh Technol Conf 4:2424–2429
8. Li W, Joshi A, Finin T (2013) CAST: Context-aware security and trust framework for mobile ad-hoc networks using policies. Distrib Parallel Databases 31(2):353–376

9. Griffiths N, Chao KM, Younas M (2006) Fuzzy trust for peer-to-peer systems. In: IEEE international conference on distributed computing systems workshops, pp 73–73
10. Xia H, Jia Z, Ju L, Li X, Sha EHM (2013) Impact of trust model on on-demand multi-path routing in mobile ad hoc networks. Comput Commun 36(9):1078 1093
11. Agrawal R, Khiani S (2013) Energy efficient and trust-based node-disjoint multipath routing protocol for WSN. Int J Sci Res 4(3):1069–1073
12. Verma J, Shukla PK, Pandey R (2016) Survey of various trust based QoS-aware routing protocol in MANET. Int J Comput Appl 137(3):34–43
13. Azzedin F, Ridha A, Rizvi A (2007) Fuzzy trust for peer-to-peer based systems. World Acad Sci, Eng Technol 27:123–127
14. Rehak M, Pechoucek M (2007) Trust modeling with context representation and generalized identities. In: International workshop on cooperative information agents. Springer, Berlin, pp 298–312
15. Rathnayake U, Sivaraman V, Boreli R (2011) Environmental context aware trust in mobile P2P networks. In: Local Computer Networks (LCN), 2011 IEEE 36th Conference on, pp 324–332
16. Cho JH, Chen R (2013) On the tradeoff between altruism and selfishness in MANET trust management. Ad Hoc Networks 11(8):2217–34
17. Geetha S, Geetha Ramani G (2014) Survey of trust-based routing protocols in MANET. Int J Adv Res Comput Sci Softw Eng 4(10):604–608
18. Parvathi Raj P, Krishnan K (2015) A novel energy-efficient routing for data intensive MANET. Int J Recent Innov Trends Comput Commun 3(8):5318–5321

Security in Ubiquitous Computing Environment: Vulnerabilities, Attacks and Defenses

C. Shoba Bindu and C. Sasikala

Abstract Ubiquitous computing is a computing paradigm, which enables computing to be appear everywhere using any device, in any location and any format. It includes resource constrained mobile and wearable devices, where computations are embedded in the environment (everyday artefacts). Those devices are connected to each other using infrastructure-based as well as mobile ad hoc networks. Due to the, resource constraints and limited internet connectivity the traditional security mechanisms such as Public Key Infrastructure (PKI) and Server centric authentication, are not used in ubiquitous computing. However, to enjoy the numerous benefits offered by this computing paradigm, we must address the security issues related to this computing. In this chapter, we discuss security issue such as location privacy, Authentication and device pairing and RFID. The major part of this chapter is intended to discuss the security challenges: Vulnerabilities, attacks and possible solutions in the Ubiquitous Computing environment.

Keywords Ubiquitous computing · Security · RFID · Location privacy

1 Introduction

After the invention of the first computer, the computing paradigm has constantly been evolving and has gone through significant changes over the past decades from Microsoft's slogan computer on every desk to multiple devices for the user. In recent years, it has led to the development of standards and technologies for wireless communications such as WLAN, RFID, LTE, NFC, and all types of mobile, embedded and wearable devices [1–3]. As a result, in recent years, the perspectives of the ubiquitous computing introduced by Mark Weiser in 1988 [4] come into true. He

C. Shoba Bindu (✉)
Department of CSE, JNTUA College of Engineering, Anantapur, India
e-mail: shobabindhu.cse@jntua.ac.in

C. Sasikala
Department of CSE, JNT University, Anantapur, India

© Springer Nature Switzerland AG 2019
N. Jeyanthi et al. (eds.), *Ubiquitous Computing and Computing Security of IoT*,
Studies in Big Data 47, https://doi.org/10.1007/978-3-030-01566-4_5

imagines the world with technologies that are not visible to the user instead of that they can be embedded into the environment. For example, smartphones, tablets or smart watches, smart contact lenses, as well as smart glasses or smart implantable medical devices, provide universal access to different types of information.

Nowadays, many people use Internet services from their home, office or any frequent places they visit. However, many users can access their emails on their mobile phones or receive their services through their Personal Digital Assistants (PDA). As computing becomes more pervasive, people are always expected to access their services and information at any time and anywhere. It leads to the development of ubiquitous computing. However, with the emergence of new technologies and devices, many novel security issues arise. In Ubiquitous Computing, security is important because there are more people than before using software-based devices as part of everyday life. Modern mobile devices are used for both personal and professional communication, and also for various types of information such as financial information, personal text messages, or sensitive information such as photos with mobile devices, etc.

The revolution has transformed society in many ways. In this chapter, we will discuss the security of Ubiquitous Computing. A modern computer is a very complex system even an expert also cannot understand it at all levels. When we consider a system of many networks instead of a standalone machine and such networking is spontaneous and not visible such as wireless technology. In addition to the complexity of this system, there are many hidden bugs and unexpected interactions that cause unpredictability and undesirable behavior. Unfortunately, such an error and interaction can be exploited by the adversary who causes internal damage. However, as networked systems are more extremely embedded in the environment, and the damage caused by malicious attacks has become more varied. Nowadays, computer security can also seriously affect people who do not use computers because they are probably using them and rely on them without knowing what they are doing.

1.1 Ubiquitous Computing

Ubiquitous Computing (UbiCom) is a computing paradigm, where computing is made to appear everywhere using any device, in any location and any format. Here, computations are embedded in the environments. With advancements in computer science and technology, the computer applications are integrated into our everyday life. The devices worked in the network, and the independent environment has the capability of communication with humans and with each other. These devices support access to context-aware applications, mobile users, location-aware services and mobile data access. It provides anytime anywhere access to the data services while making the presence of the system invisible to the user.

Three main properties of Ubiquitous computing are [5], distributed computation, context-awareness, and invisibility. The distributed computation refers to the computers/systems need to be networked, distributed and transparently accessible, they

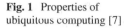
Fig. 1 Properties of ubiquitous computing [7]

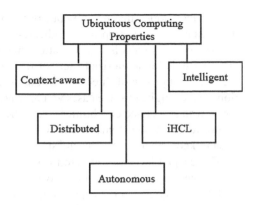

can interact with each other as well as with humans. The context-awareness property refers to optimization of system operation in their physical and human environment, it is necessary to make the systems aware of environmental context. The invisibility property makes the computer interaction with humans should be more hidden. However, the objective ubiquitous computing [6] is to be user-centric that allows users to interact with the system in a natural and non-intrusive way to assist users in everyday life. The five core properties of the ubiquitous computing are illustrated Fig. 1. It is not mandatory for ubiquitous computing solutions to satisfy all five properties but in practice it may support all five properties.

1.2 Security

Security is the major obstacle to adopt any new computing paradigms. Traditionally, the threats to information systems can be classified concerning three fundamental security properties that they are: confidentiality, which means only authorized users can read the information. Integrity, it ensures only authorized users can modify the information and availability mean ensuring that only authorized users can access and use the system as intended, without any delays. These are a good starting point to understand what's most important, even though they do not necessarily exhaust all the desirable security properties of an information system. Generally in many cases people automatically assume that the primary security concern is the protection of confidentiality which means unauthorized users shouldn't be able to see private data. But, integrity is instead more significant that is unauthorized users must not be able to edit the user privatedata. And the primary concern is about availability, especially when the system integrates several different functions, for example the college computer system cannot stop working for more than a week, as it handles not only the students' grades but also the staff payroll, the course timetable, and the library catalog.

You may notice that the previous paragraph has a recurrence of the phrase "authorized users". It appropriately indicates that the foundation upon which the three security objective rest is authentication, means it makes the distinguishing between authorized and unauthorized users of the system. Usually, this foundation usually includes a sequence of three sub-actions: identification, verification and authorization. Interestingly, in some cases you may not recognize a particular person, but just a role with associated capabilities—e.g., the librarian checks an authorization token to ensure that this customer is entitled to borrow up to four books, but she does not need to know about her name and surname to grant her volumes.

This chapter is organized as follows: Sect. 2 explore the security issues in Ubiquitous computing such as Location privacy, Scct. 3 discuss the security issues related to RFID Technology, Sect. 4 explains the fundamental security issues in Ubicom such as authentication, confidentiality, integrity and availability. Finally, Sect. 5 explains the Security challenges: vulnerabilities, attacks and Defenses.

2   Security Issues in Ubiquitous Computing

The Ubiquitous computing model uses wireless and mobile infrastructure. Since this architecture plays a major role in how users can interact with ubiquitous services and applications. There are many issues related to mobile and wireless infrastructure in the ubiquitous computing environment. We think that more research is necessary for this area for deriving the specific role and requirements. The requirements could include a universal or integrated access to various wireless and mobile networks, support for reliable communications, group communications (multicast), and interworking of these technologies. Many of these issues are not resolved and we believe that they should be covered before the vision of ubiquitous computing come into reality.

Security is the major issue in ubiquitous computing, as individual, organizations or groups, are unlikely put their important, personal and mission-critical information over an infrastructure that is either not secure or is not identified to be secure. The security weaknesses of wireless and mobile infrastructure stem from both, due to the implicit weaknesses in certain wireless security algorithms such as wireless LANs, and the use of several "incompatible" security schemes. Because of the several reason's strong security has not been implemented in wireless infrastructure. There are many security issues in the ubiquitous environment, including confidentiality, integrity, authentication, authorization, accessibility and non-repudiation. Other issues would include convenience, ease-of-use, speed and standardization. So, a security strategy must be designed and implemented based on the type of data and the cost of possible loss, modification, and stolen data. In addition to security and privacy risks, new vulnerabilities arise due to the use of wireless devices. The wireless infrastructure may have several wireless networks with different levels of security. These can lead to possible change or deletion of information, and denial of service. In addition to these, many more security issues arise due to feature interactions, poor

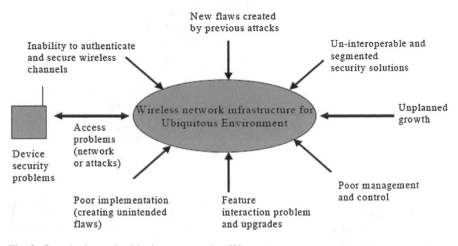

Fig. 2 Security issues in ubiquitous computing [8]

implementation and unplanned growth and new flaws that are created due to prior attacks as shown in Fig. 2.

2.1 Location Privacy

One of the buzzwords of ubiquitous computing is "location-based services." Ubiquitous computing environments, provides location-specific services based on the availability of people location information. However, the location is a sensitive information and giving it to the random entities might pose security and privacy risks [9]. For example, to limit the risk of being robbed, people want to keep their location information secret when walking home at night. Therefore, only authorized entities should have access to people location information. While the location information has gained much attention, its access control requirements are not fully studied thoroughly. Location information is completely different from information such as files stored in a file system where access control requirements are studied widely. So, Location information is different because there is no single point at which access needs to be controlled. Instead, a variety of sources such as a personal calendar or a GPS device can provide the location information. Also, different types of questions can provide the same information. Therefore, a system that provides location information must perform access control in a distributed way by considering different services and interactions between queries.

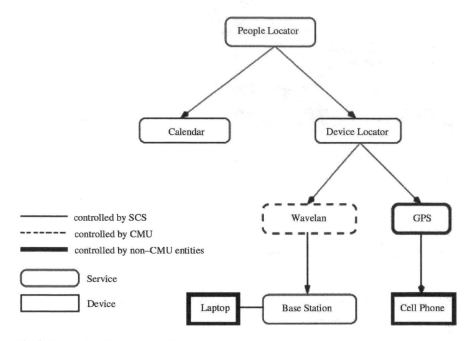

Fig. 3 People location system [10]

2.1.1 People Location System

The hierarchical structure of people location system shown in Fig. 3. The nodes in the graph represent either services or devices and the arrows indicates which service interacts with which other service or device. The location system is the combination of multiple location services. Each location service uses either a particular technology for gathering location information or processes location information received from other location services. Usually, the flow of location information is in reverse direction of a request. To provide efficient scalable and robust services, a location service can be implemented either on a single or multiple hosts. These location services can be divided into two groups, one group includes the services that are aware of the people location information such as People Locator service, Device Locator service and Calendar service. The second group has the services, those are aware of the location of devices. These services identify the user indirectly by identifying the device(s) that the user is carrying with her.

The People Locator service aggregates information received from other services. The Calendar service determines their current location based on people's appointments. To know the location information of a person the Device Locator service maps a query to possibly several queries of her devices and contacts corresponding services in the second group. In the Fig. 3, this group of services consists of GPS service and the Wavelan service. The GPS service retrieves the location from GPS enhanced mobile phones. The Wavelan service keeps track of the location of wireless devices by identifying their base station

2.1.2 Location Privacy Issues

Location privacy issues arise not just with mobile phones but to a certain extent with any mobile device that communicates with fixed infrastructure while on the move. The primary way to protect location privacy is through access control by giving you the power to grant or deny access to your location information based on the identity of the requester, as well as other features of the specific query being made. If we follow this strategy, an application will know that someone is there in that place, and it will provide the service to that person, but no one knows who it is. Even though this protection is good, it is not hard to find out which person is present at that location other than the owner of that location. Sometimes simple heuristics approaches [11] are also used to resolve the ambiguity.

A Countermeasure for the above problem is "mix zone" [12], it is an unobservable spatial region in which we prevent the adversary application from tracking us. We hope that, whenever we leave the mix zone, the application will be mixed up (i.e. confused us) with someone else. Once again we then take on the role of the adversary application and try to break the anonymity by providing a correlation between those who entered into the mix zone with those who came out of it. Some researchers also developed quantitative measures for location privacy. They allow us to assess the effect of any privacy protecting countermeasures we may adopt objectively. For example, degrading the spatial or temporal resolution of the data provided to the applications. To ensure a certain level of anonymity for the entities being located, Marco Gruteser et al. [13] proposed an alternative safeguard to degrade the spatial and temporal resolution of the answers from the location server to the applications.

Jackson et al. [14] proposed one of the most effective ways of protecting location privacy is to reverse the architecture, that is infrastructure tell the mobile nodes where they are (as happens in GPS) instead of having mobile nodes (e.g. tags or cellphones) that transmit their location to the infrastructure. For example, instead of the user asking which restaurants are there nearby, the restaurants would broadcast messages with their location and the type of food they offer, which mobile nodes would pick up or discard based on a local filter set by the user's query. Scalability can be achieved by adjusting the size of the broadcast cells (smaller, more local cells to accommodate more broadcasters and more frequent broadcasts). Note that in this reversed architecture the restaurants simply cannot know who receives their advertisements (hence absolute privacy protection), but in the previous case the server knows and can inform them. It can be a financial incentive against the deployment of such an architecture.

George Daneziset.al [15] and others have studied the Location privacy problem from an interestingly different perspective, with the aim of highlighting how much users value their privacy. Their studies use experimental psychology and economics techniques to extract from each user a quantitative monetary value for one month's worth of their location data (median answer: under 50 EUR), then they compare it with national groups, gender and technical awareness.

2.1.3   Location Policies

Location queries can arise both from people and from services. A query can ask for the location of a user or the people in or at a geographical location, such as a room in a building (room query). Based on the two primary queries, it is possible to create more sophisticated queries or services that provide location-specific information. To prevent location information from unauthorized entities, we may enforce location policies. Location policies allow an entity to access location information about a person or the people in a room, only if permitted by that person's and that room's location policy respectively. In this section, we examine some location policies and present requirements that need to be provided by the access control mechanism of a people location system.

User and Random Policies

Corresponding to the two primary queries, there are two types of location policies: user policies and room policies. A user policy specifies who is allowed to get location information about a user. "For example, Bob is allowed to find the location of Alice". A room policy states who is permitted to find out about the people in a room. "For example, Bob is allowed to find out about the people in Alice's office". Also, both user and room policies should be able to limit information flow in other ways. We believe that at least the following properties should be controllable:

Granularity: A policy can restrict the granularity of the returned location information. For example, a user policy can specify the building in which a queried user is present is returned instead of the exact room (e.g., \CMU Wean Hall" vs. \CMU Wean Hall 8220"). A room policy can state the number of people in a room is returned instead of the identity of the person in the room (e.g., \two people" vs. \Alice and Bob").

Locations/users: User policies includes a set of locations (e.g., buildings or rooms). If the queried user is at one of the listed locations then only the location system will return the location information. For example, Bob is allowed to know about Alice's location if she is in her office then only he identifies her location". Similarly, room policies can include a set of users. The answer to a room query will include only users listed in the policy (provided they are in the room).

Time intervals: Location policies may be limited to certain time intervals during which only access should be granted. For example, access hours can be restricted to working hours.

In previous work by Spreitzer et al. [16], users can set the boundaries on room policies such as users must specify whether they want to be included in results to room queries or not. This approach is appropriate for some scenarios only not for all. Some people think that, regardless of the user policies of the users in her room, the owner of a room should always be able to find out who is in her room.

User versus Institutional Policies

Depending on the environment, different entities specify location policies. For some environments, a central authority defines policies, whereas for others, users set them. Also, some environments might give both users and a central authority the option to specify policies. In general, governments and companies probably do not want the location of their employees or the people in their buildings to be known to outsiders, whereas this information can be delivered to (some) entities within the organization. In such cases, a central authority would establish the location policies such that no information is leaked. For other environments, such as a university or a shopping mall, the institution behind the environment cares less about where an individual is or who is in a room. For these cases, it should be up to an individual to specify her user policy. We examine some example environments in more detail in the following section.

Transitivity of Access Rights

If Bob is allowed to access Alice's location information, should he be allowed to send this access to Carol? If Ed is given the right to know people in his office, should he be allowed to grant this right to Fred? In short, can access rights to location information be transitive? The answer to this question depends on the environment. The policymakers must specify whether they want access rights to be transitive or not. Even if a user does not allow other users to forward his access rights, he may still issue queries on their behalf. One way to deal with this problem is to take the access right back from the user.

Conflicting Policies

User and room policies can conflict. For example, assume that Alice does not allow Bob to locate her, but Carol allows Bob to find people in her office. If Alice is in Carol's office, should the location system tell Bob about it? There are many ways to deal with this problem: The Location system ignores the room policy when answering a user query. Similarly, it ignores the user policies for a room query. In our example, if you ask people at Carol's office, Bob sees Alice, but he is not allowed to ask a user query for Alice. The system looks at both policies for any request and returns information only if it is approved by both of them. Thus Bob should not be able to see Alice being in Carol's office. The user and room policies are fixed in a synchronized fashion so that no conflicts arise. For example, Leonhardt and Magee [17] suggested authorizing user/room pairs, Alice and Carol's location policies would thus have to be rewritten. The efficiency of this approach depends on the environment in which the location system is deployed.

2.1.4   Example Environments

In this section, we explore how location policies are specified and applied in two different environments; a hospital and a university environment.

Hospital

Medical information, such as patient information, is protected by a multilateral security model [18], which protects the flow of information between the compartments. For example, doctors who care about a patient have her medical data, but not every doctor in the hospital. A similar model is required for location information. The patient's doctor only need to locate her. Also, a patient must be able to allow other people (e.g., her husband) locate her. To meet these requirements, the hospital will set up central authority policies. This can give her the right to include additional people in the user's policy. Room policies should be established through the central authority to protect the privacy of patients. The hospital scenario does not need to synchronize user and room systems. Patients may be allowed to specify transitive user policies, but doctors cannot forward an access right given to them in a user policy. Room policies should not be transitive.

University

In the university setting, students and faculty members derive their customer policies. However, the room procedures were established by a central authority. In an office, the authority office is likely to give the right to establish its room policy to the member of the office. For lecture rooms and hallways, the authority will set the room policies such that user and room policies need to be synchronized. That means, when the room receives a query, the location system asks the user policies in the room/hall before the returning their identity. In this scenario, user and room policies are not synchronized for office. Here, User policies may be transitive but room policies may not be transitive.

3   RFID Technology

Radio Frequency Identification (RFID) has recently received a lot of attention as a growing technology in the ubiquitous computing Environment. Because it does not require line-of-sight alignment, it identifies the multiple tags simultaneously, and it tags does not destroy the integrity of the original object. It has many benefits like the low cost of passive RFID tags and the fact that it works without a battery. RFID systems [19] have two main components: the RFID tag, which is attached to the object to be identified and it serves as the data carrier, and the RFID reader, which can read from and sometimes even it writes the data to the tag. The tags typically contain a microchip that stores data and a coupling element, such as a coiled antenna that is used to communicate via radio frequency communication. The readers usually

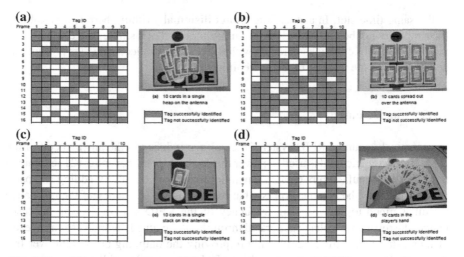

Fig. 4 Four different arrangements of 10 playing cards equipped with RFID tags: **a** in a heap on the antenna, **b** spread out over the antenna, **c** stacked on top of each other, and **d** in the player's hands. The patterns to the left of the images show a snapshot of the data captured by the reader. A dark field indicates a successful detection of a tag in a frame; a light field indicates a failed detection. These measurements were carried out with 32-time slots per frame and a frame rate of 5 Hz [20]

include a control unit, a radio frequency module, and a coupling element to interrogate the tags via radio frequency communication.

3.1 Failed RFID Tag Reads and Their Causes

Failure to detect RFID tags that are present in the read range of a reader can be due to a variety of reasons including tag detuning, collisions between the tags, air interface, tag misalignment, and metal and water in the vicinity of the RFID system. Failed tag reads caused by some of these phenomena is illustrated by Romer et al. [20] by taking playing cards with RFID tags. In addition to this some other scenarios also used to demonstrate some of the challenges involved in RFID, they believed that the playing card scenario is an ideal example because it fairly demonstrates the most common causes of failed tag reads. Fig. 4 shows the RFID tags on the back of the playing cards with the RFID antenna of the I-CODE System in the background.

3.1.1 Tag Collisions

In most cases, tags that do not transfer their ID at the same time slot are not recognized. Exceptions to this rule are due to the capture effect [21], where the reader manages to properly identify the data sent with one of the tags, even though many tags react at the same time slot. In a stochastic anti-collision algorithm, there is a possibility for a tag may not be recognized for a minimum period of a single frame. Obviously, the probability of collisionsincreases with the number of electronic tags present and decreases with the number of available time slots.

3.1.2 Tag Detuning

In inductively coupled RFID systems the voltage induced in the antenna coil of the tag by the magnetic field is used to power the microchip. Finkenzeller [21] explained how the tag manufacturers created a parallel resonance circuit by adding a capacitor in parallel to the antenna coil, so that the resonance frequency of the resonance circuit is tuned to the operating frequency of the RFID system. During the resonance, the resonance voltage generated throughout the tuned tag increases as a result of increasing readability, significantly improving the outgoing bandwidth over frequencies. As a resonant application, the tag is, however, vulnerable to environmental detuning impacts which can also cause a significant reduction in reading distance. For example, a group of RFID tags that are close to each other, exhibit significant detuning effects caused by their mutual inductances. Undesirable changes in the tag's parasitic capacitance and effective inductance can also occur by metal and different dielectric mediums in the vicinity, e.g. a hand holding the tag. The change in resonance frequency away from the operating frequency results in the tag receiving less energy from the reader field and hence a decrease in reading distance. Tag detuning due to other tags that are very close it also cause the low read rates.

3.1.3 Other Sources of Error

Other factors of failed reads include the presence of metal in the tag vicinity environments because it distorts the magnetic flux, thus weakening the energy coupling to the tag. If the tags are attached to a metal surface, they can often not be detected at all. Similar to tag detuning, due to the antenna detuning metal in the vicinity of the reader antenna results in a read range reduction. Failed reads also caused by the misalignment of the tags with the magnetic field of the reader coil. Maximum power transfer occurs when the tag coil plane is perpendicular to the magnetic field lines. As the label is rotated concerning the field lines, the coupling is reduced until the tag is no longer identified.

3.2 Security and Privacy

To handle multiple tags reading securely and reliably, some security techniques have been developed [22], i.e. Session Concept, Enhanced Secured Protocols, Ghost Reads Improvements, Dense Readers Conditions and Covered Coding. However, there are still some privacy concerns in both user and application levels. They are also facing some security loophole in defining and managing the 'random number' key that requires attention to avoid possible privacy violations or information leakage by eavesdropping on the communication channels.

3.2.1 Fake Tag ID Problems

Generation 2 RFID standards use various implementations to make sure that the incoming tag ID is in fact a valid tag ID rather than noise or glitches(ghost read). Ghost read means, a signal processor interprets noise to be a tag IDand it was a major obstacle in adopting this valuable technology. One of the drawbacks of Generation 1 RFID protocols is ghost read. As a solution to this problem Generation2 RFID protocols come up with a 'Query' Concept. Communication between tag and reader is defined with timing constraints to create an illusion of a full duplex link. In practice, the communication is still maintained in a half-duplex mode. The tag does not speak when it reads the reader commands, but the time limits respond within a predefined time. If the tag fails to respond within the given time, the task will be terminated and the whole process must start from the beginning.

3.2.2 Password Protection and Effective Randomness

Maintaining a secure link between reader and tag is crucial for preventing data transmitted over an air interface. In Generation 1:class 1 standards of RFID, an 8-bit password is used to execute the 'kill' command to secure the data. This 8-bit password is neither secure nor hard to break because of just 256 possible values for the password. In Generation 1 Class 0, a 24-bit password is used, which provides better protection for accessing the data. Generation 2 uses a 32-bit password, so it provides 4 billion possible values to search for the correct password. Thus, RFID achieved a level of secure communication that it never before [23].

In Generation 2, along with the password concept, a random number is used to scramble data. The tags will generate and uses a 16-bit Random Number Generator (RNG) throughout the communication session due to this we can ensure that the communication link is safe. For example, having a tag of the population up to 10,000, then the probability of generating the same sequence at the same time is less than 0.1%. In addition to 32-bit password protection, Generation 2 RFID standards use cover coding while disclosing the data with a random number to randomize the data [24].

3.2.3 Security Problems and Possible Attacks

As a technology with a convergence tendency, the RFID reader can be integrated into a hand held devices or mobile phones. These Low-cost tags led to widespread adoption of the technology and deployment on such huge scale, creates new threats to users and application privacy due to the powerful tracking capability of the tags [25]. All UHF standards provide a security mechanism for reading user memory but any reader can read the tag ID on the fly [26]. A security check should be performed on the tag before the ID is transferred and a mechanism must be defined to identify a credible reader to resolve the privacy issues.

The transmission protocol of tag reader and reader to tag communication defines the process of exchanging the data and instructions between the reader and the tag in both directions. This protocol works based on the concept of "reader talks first" and means that every tag according to UHF standards will always answer to the reader's query with its identification (ID) at first. It makes the technology powerful and an intruder can track the reader tag. The attacker can obtain concrete product information associated with the EPC/UID code, it is available on the public network. Even though the current UHF protocols have a 'kill' command, the tag is currently dead and it will be implemented before moving it to the end users' hands, but this is not a solution for most applications. Some applications such as vehicle tracking system and personal identification systems require tags associated with objects for security purposes. Another security issue related to the tag is specified as, here the communication between a tag and a tag reader is using radio frequency, and so anyone can access the tag and obtain its output. Hence, there is a possibility for an attacker to eavesdrop on the communication channel between tags and readers. Therefore, the authentication scheme used in RFID system must be able to protect the data passing between the tag and the reader. It means the scheme itself should provide some encryption capability [25].

Generation 2 RFID standards provide a good mechanism to transfer data securely between the tag and reader. The exchange of cover-coding was first initiated by a random number request, i.e. RN16, from the Tag. If lower secured plaintext or mechanisms are used, eavesdropping on the communication channel may break the entire security process of the cover-coding. The generation and management of this 'random number' are important for ensuring the security and integrity of the system but its size should be reconsidered and the time of command to response should be limited with precise values. So, that the random number is directly proportional to the time for the command to response. Even though, the random RN16 connection provides secure communication link, its 16-bit size still makes it susceptible, as generating 65,536 combinations is very easy to find d out those combinations even with simple processors. Also, the duration of the command to response time makes it more vulnerable, it means that reader A would start querying the tag but reader B (an intruder) can join in the communication link with a fake random number.

3.2.4   Possible Solutions

To break the cryptosystem, any cryptanalyst must use extremely large amounts of computing recourses and time to analyze the data even if he knows the whole crypt-analytic process. One of the requirements of this cryptosystem is the property of changing few parameters that result large change of the whole system. For RFID, the cryptosystem must be easily implemented, so floating-point operations and other complicated numerical operations should be avoided.

4 Authentication and Device Pairing

Ubiquitous computing and Peer-to-peer systems involve many principals, but their network connectivity is irregular and not guaranteed. Traditional methods for authentication [27, 28], from Kerberos [29] to public-key certificates, are therefore do not work, because they are based on online connectivity to an authentication or revocation server. We need new solutions.

Secure transient association

The main application of authentication to intermittently connected networks is a secure transient association. There are many examples of this paradigm of interaction in applications such as mobile computing, medical equipment, consumer electronics, automatic teller machines, car security systems and weapons systems. To visualize secure transient association, consider the following scenario: In the ubiquitous computing world, you no longer want to litter your coffee table with some remote controls for your TV, DVD, stereo, VCR, curtains, air conditioning, central heating. Instead, you want all of these systems to obey a universal remote control, which for the sake of argument will be some PDA. Because you no longer buy the remote control with the appliance, you need to be able to establish an *association* between the two after purchasing the appliance. You want this association to be *secure*, because you don't want your neighbor to be able to activate your appliances (whether by accident or malice). And, if you want to resell your old DVD while keeping your PDA, and you want to replace a broken PDA without losing control of all your appliances, you also want this association to be *transient*, or revocable.

4.1 Resurrecting Duckling

The Resurrecting Duckling security policy model Stajano et al. [30] was developed to solve the above problem and in particular to implement secure transient association: you want to bind a slave device (your new flat screen TV) to a master device (your cellphone used as a universal remote controller) in a secure way, so that your neighbor can't turn your TV on and off by mistake (or to annoy you) and so that the stolen TV is useless because it doesn't respond to any other remote controller; but also in a transient way, so that you can resell it without also having to give the buyer your cellphone. It is based on four principles proposed by Stajano [31]: It describes the way of establishing asecure transient association between a master and slave. Following four principles define the Resurrecting Duckling.

1. **Two State principle**. The entity that the policy protects, called the duckling, can be in one of two states: imprintable or imprinted. In the imprintable state, anyone can take it over. In the imprinted state, it obeys only its mother duck.
2. **Imprinting principle**. The transition from imprintable to imprinted, known as imprinting, happens when a principal (mother duck), sends an imprinting key to the duckling. It must be done using a channel whose confidentiality and integrity are adequately protected (physical contact is recommended). As part of the transaction, the mother duck must also create an appropriate backup of the imprinting key.
3. **Death principle**. The transition from imprinted to imprintable is known as death. It may only occur under a very specific circumstance, defined by the particular variant of the Resurrecting Duckling policy model. So, one has to chooseone among the following.

   - The order of the death starts with the mother duck (default).
   - If a predefined time interval expires then death by old age.
   - Death after completion of a specific transaction.

4. **Assassination principle**. The duckling must be built in such a way that it will be uneconomical for an attacker to kill it, i.e. to cause the duckling's death artificially in thecircumstances other than the one prescribed by the Death principle of the policy.

Note that the Assassination principle implies that a duckling-compliant device must be endowed with some appropriate amount of tamper resistance. The Resurrecting Duckling policy has very general applicability. It is not patented and therefore it has been applied in a variety of commercial situations. It has been extended by Stajano [32] to allow the mother duck to delegate some or all of its powers to another designated master.

4.2 Confidentiality

When people think about security issues for ubiquitous computing, they first think of eavesdropping as a consequence of wireless networking. But this concern is changed: once we have addressed the critical issue of authenticating the principals and sharing key material, we have mature and robust symmetric ciphers for protecting a communications channel's confidentiality. The actual problems are elsewhere.

Bits per second or Bits per joule?

The size and shape of the typical ubiquitous computing device impose new constraints. Untethered devices are battery powered, so they cannot use the fastest and most powerful processors available. To avoid that problem they require very frequent recharges (as laptop users know all too well). Many ubiquitous computing devices therefore have processors that are too slow for computationally intensive tasks such as public-key cryptography.It is well known, and one traditional way of dealing with those processors is to do most of the work as background tasks or as pre computations. But the batteries of small portable devices hold only a small, finite amount of energy; this places a bound on the total amount of computation the devices can perform, rather than on the rate at which they can perform it. This problem is new and more interesting: to evaluate a cipher (or any other algorithm) on a device, the most relevant performance figure is no longer bits per second, but bits per joule.

Biometrics, Coercion, Traffic analysis, and more

While it is straightforward to protect the confidentiality of wireless traffic, it is very difficult to protect the privacy of the information in the devices. At present, some people are concerned about this. Most of the PDAs, for example, which is relatively likely to be stolen or lost, are not even password protected. Even though those are password protected, they do not use encrypted storage. Therefore, those devices are moderately at the mercy of the effective attacker. It is hardly surprising for both owner and thief, even the value of the information held in the typical PDA is small.

In the future, however, the ubiquity of computing devices will multiply the opportunities for storage of information about our activities. Our digital butler's mission is to have the task of identifying and remembering as much as possible about our habits and behavior. Finally, the technology infrastructure introduced for one purpose has once again misused, think about credit cards being used for purchasing. It is important to protect the confidentiality of the data held in at least the mother-duck devices. There are three components to address this problem. The first thing, is to find techniques that allow users to authenticate their devices, such as a password or biometric (such as a manuscript signature with a pen in PDA) and much more difficult than it looks. Secondly, protects the keys within the device such as any long-range keys that are used to encrypt personal data(user's profile), and imprining keys of the controlled devices. Third, Security Recovery—The problem of recovering from a PDA theft when the device is live or when the thief guesses the owner's password The fourth is the problem of resistance to the coercion.

Finally, we need to consider metadata protection. Anonymity, traceability, and traffic analysis are aspects of confidentiality, so far that hasbeen underestimated, but they will take more prominence in the ubiquitous computing. Encryption makes it easy to protect a conversation; but the terms when, from, and to are not to mention the very fact that a conversation is taking place remain observable. Defending against traffic analysis is a difficult problem, and an active research area. From the user's perspective, it is necessary to support location privacy and make the task of combining user's transactions with each other is difficult, it is necessary at the design stage; otherwise the ubiquitous computing infrastructure will become a tool for ubiquitous surveillance.

4.3   Integrity

The basic integrity problem is to ensure that messages from one party to another are not corrupted by a malicious third party. This is similar to confidentiality in that, once we know how to do authentication and key distribution, the problem is trivial to solve using well understood cryptographic mechanisms, such as message authentication codes. Authenticating broadcast data is somewhat trickier if we wish to avoid the power cost of computing a series of digital signatures, but researchers have devised several chaining protocols to tackle this problem. The most serious integrity problem for ubiquitous computing, therefore, is once again not with the messages in transit but with the device itself.

Tamper resistance and Tamper evidence

How do I know whether a device I am using for communication has been modified or even replaced with a duplicate? It is easy to identify this as an authentication problem, and there is a close relationship between integrity and authentication. To address this issue, Duckling solution may be helpful. However, there is, another aspect to the solution—physical tamper protection.

The usual assumption underlying authentication is that the network is insecure and it is controlled by the attacker, but that the users involved have the ability to keep their secrets. Ubiquitous computing takes this assumption and, as it does with so many others, turns it on its head: network attacks are often easier to deal with, but the attackers destroy many principals in the network. Providing high-level tamper resistance, which makes it impossible for an attacker to access or modify the secrets held inside a device, which is very expensive and difficult. It is often better to rely on temper resistance instead of tamper evidence stated by Anderson et al. [33]. It ensures that tampering attacks and traces of attacks. The main objection to this strategy is that it breaks the loop of machine-based verification. A physical seal's integrity cannot be verifiable as part of the authentication protocol; instead, it is necessary to inspect human. Some might see this as a security hole, but it could be an advantage. It means that the responsibility for protection depends on the person relying on that protection, rather than with some third party who might have different motives. Moreover, managing the protection is also a matter of common sense. The two common reasons for security failure are that the principal responsible for the security is not the principal relying on it, and that technical approaches such as public- key certification is difficult to understand and manage. It is somewhat certain for engineers to assume that they can solve all problems using mechanisms within their realm of professional expertise, when other, simpler, mechanisms might be more robust.

4.4 Availability

The traditional attack on the availability of the wireless system is to jam the communication channel. The Ubiquitous systems that rely on short-range RF communication can fail in the presence of jamming, but the techniques that deal with it lie outside system design. Once the jammer goes out of the range (or once the police take him away), the network can start normal operation. And the possibility of a Denial-of-service attack arises from the relationship between security and energy conservation. If a device has limited battery energy then it tries to sleep as much as possible to conserve it, keeping it awake until this energy runs out can be an effective and selective attack. After the battery is flat, the attacker can walkaway, leaving the victim disabled. We can call this cruel treatment as sleep deprivation violence.

You might think that authentication could prevent such attacks, but this is not always the case. Authentication lets you distinguish friends from unknowns. But in some applications you cannot refuse to provide services to unknowns. For example, if you are a Web server. The problem for the server is whether to answer queries from unknowns: they might be planninga denial-of service attack, but they might be genuinely interested in the answer. So, Identifying repeat specters is good because they can easily fake the source information and they can also corrupt multiple "innocent" principals into cooperating in the attack this is called DDOS (distributed denial-of-service) attack.

When the server has many functions of different importance, they can be prioritized and it follows the resources allocation strategy to tighten resources to less important uses. This guarantees a particular level of service to the more important uses. In fact, it still fails to protect against certain types of attacks, such as those from authorized insiders. One approach to this problem is the plutocratic access control: if you got the money to pay you will receive the service. Charging for accessibility limits how much to keep clients indiscriminately asking for resources. In fact, if the server is making a profit in the customer service, denial of service problem is no longer concern, which means that the server can make more money! If it is not practical to collect the original money, the server can still use the same limited strategy to get some expensive sacrificial ritual in exchange for service. Many researchers have suggested that clients can solve cryptographic puzzles or answer a question that is easy for a man but hard for a machine. This approach is more suitable for peer-to-peer applications, but the former solution is better for ubiquitous computing environments.

5 Security Challenges in Ubiquitous Computing: Vulnerabilities, Attacks and Some Defenses

5.1 Vulnerabilities

Network Dynamics: Ubiquitous computing environment has the characteristic for its absence of a fixed infrastructure and a lack of a central server, central authority, and centralized trusted a third party. As mobile devices may join or leave the network at any given time, it is very important for a network to self-configure.

A large number of nodes: The second most commonly addressed challenge results from the large number of nodes participating in network communications in a ubiquitous computing environment. Since some nodes may act selfishly (refuse to forward packets to other nodes), maliciously (seek to damage network operations) or show signs of a dynamic personality (behave strategically in a way that best benefits them). If we do not handle carefully it may lead to a collapse of a whole network. So, in trust computation and management handling bad behavior nodes is a challenge.

Resource constraints: In ubiquitous computing environment nodes are heterogeneous devices that differ in their battery-life, processing power and communication capabilities. Due to the devices' resource constraints, ensuring the availability of services and design security mechanisms that rely on complex computations is a challenging task.

Authentication-related challenges: The goal of ubiquitous computing is to design environments in which people could not be aware of the technologies (embedded in the environment) surrounding them. Existing authentication mechanisms still do not entirely adaptable with this vision. For example, we have mobile phone locking mechanisms, such as pattern-drawing or passwords these are necessary to protect the user's sensitive data from the unauthorized access. But, they still require device owner's attention to prevent them from their surroundings. Moreover, by having more devices, users may face challenges in remembering all the passwords. As a result, some of the users use the same password for multiple devices or some of them store list of passwords on their device, so that the user gives an opportunity to the opponent to access the private information. Moreover, some users entirely disable their locking mechanisms, leaving their devices vulnerable.

Other vulnerabilities. Some of the vulnerabilities in the Ubiquitous environment may arise due to the lack of knowledge of services, Bluetooth vulnerabilities, frequent change of a user's context, and installation of untrusted mobile apps.

5.2 Security Attacks

In this section we discuss Some security attacks in a ubiquitous computing environment (Fig. 5).

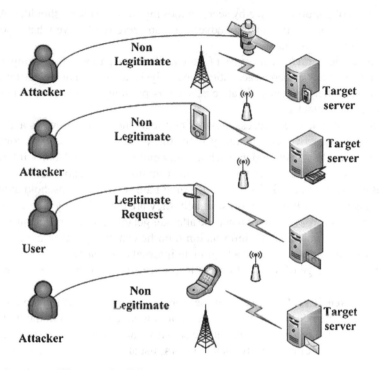

Fig. 5 Attacks on different nodes [34]

Man-in-the-middle attack: In a Ubiquitous computing environment, device authentication is important in delivering services. A user must authenticate the artifact mandatorily, while using a secret, i.e., password or PIN code. The Man-in-the-middle attack occurs when artifacts or users forward the challenges and responses to simulate the existence of other actors. When a client uses his credit card through the terminal, even though there exist proper security protocols, there is a probability of impersonation attack. In other words, an attacker can modify the transaction without having the card and without tampering with the terminal. Hence this type of attack cannot be validated in any way without a virtual context in real terminals. This kind of attack allows the impersonation of artifacts and users.

Denial Of Service (DOS): This type of attack aims at making the services and resources unavailable to its intended users. For example, by blocking the network, degrading the services, exhausting the device's resources or providing false trust ratings to the nodes participating in communication. So, it is important to take appropriate steps to find these conditions in advance.

Eavesdropping attack: This type of an attack refers to a group of attacks in which a malicious user monitors conversation to get the knowledge about confidential data or interferes with the communication channel by modifying messages. One example of such an attack is shoulder surfing, in which an adversary observes the contents on a screen of a mobile device by secretly looking over the user's shoulder. Another example is snooping, in which an adversary can secretly observe what a person is typing on his/her computer.

Cryptanalytic attack: This type of attacks include password cracking attacks, side channel attacks, such as acoustic cryptanalysis key search and electromagnetic attacks, and other cryptanalytic attacks, such as preimage, ciphertext, birthday and key generation attacks.

Access network attack: Access network connects the home gateways and outside service provider networks. Obviously, if an attacker at the home network connection point, gathers the sensitive data such as financial data, user ID other information from the network packet then the critical information can be exposed.

Illegal connection attack: Through the home gateway many household appliances are connected to multiple networks, usually they are controlled by web-based management but the problem arises when the attacker get this administration information. Once an attacker obtains this information then he can easily attack the rest of network. Also, the attackers act as a legitimate internal customer and control the home appliances. Leakage of information can lead to misuse of it that is not interested in users.

Capturing sensitive data: Electronic sensors are commonly used in the ubiquitous environment and because of their poor computational capabilities in the monitoring procedure, while an attacker can get the sensitive data from the sensor directly by putting a receiver close to it. In these sensors, usually focus is about sensing tasks instead of cryptographic affairs.

Stealing Intermediary device: Usually, an intermediary device will collect the sensor data. When the device goes into the hands of an attacker, the device cannot be reused, where it is counted as a breached source for network information to an attacking purpose. In many cases, this forms a potential vulnerability because the device contains a maintenance interface.

Data manipulation: Due to poor computational resources on sensors, they cannot directly authenticate the passing data and they use an intermediate device to store record logs of the traversing sensor. Even though encryption and decryption techniques are used for data authentication, still there is a possibility for data manipulation.

Impersonating and insiders: In this type of attack devices can be replaced with fake ones by the attacker. So that an impersonating attacker can get the sensitive data form the network.

5.3 Defenses

Security is the primary issue for the adoption of ubiquitous computing. In this type of environments and networks, to address the probable security issues, in this section we discuss some solutions.

5.3.1   Real-Time Intrusion Detection

The available intrusion detection system (IDS), is not directly deployable into the ubiquitous networks due to lack of considerations about flexibility, heterogeneity and resource limitation of ubiquitous networks. As a solution to this problem a service-oriented and user-centric intrusion detection system (SUIDS) is suggested which record events and logs to imply protection mechanisms on different network appliances against intrusions. The user-centric approach has been proposed to provide spontaneous protection against malicious users. Potential distributions are represented in the SUDIS behavior of users over a long period, which displays the desired result and relationship of any actions for a user. In brief, SUIDS includes the following:

- It accumulates the long-term behaviors of users.
- It provides the possible distributions for services.
- The normal and current behavior of users is achieved.
- It computes the Statistical deviation between established behaviors and current behavior.
- It recognizes whether the behavior is an intrusion or not.

5.3.2   RFID Based Authentication Protocol

A radio frequency Identification (RFID) is a microchip that can broadcast a unique serial number and other additional data through RF (radio frequency) signals. The purpose of RFID is to identify objects remotely by embedding tags into the objects. RFID tags are useful tools in inventory control, manufacturing, supply chain management, etc. In a ubiquitous computing environment, RFID components or RFID systems can exist anywhere. Usually, RFID tag ID State holds two values i.e. either dynamic or static. If the tag ID State is dynamic means, the tag only communicates with a fixed back-end database and if it is static means, it can communicate with any reader in a ubiquitous computing environment. RFID system must be formed to be secure against attacks such as message interception eavesdropping, impersonation, i.e., spoofing and replay and traffic analysis. Even though, RFID technology is well-suited for connecting the physical and virtual world, but before it could become a truly ubiquitous technology it still facing the challenges such as security, privacy, health and safety and creative challenges as well as technical challenges such as input data errors and system failures.

5.3.3   Role-Based Access Control

Role Based Access Control system (RBAC), is based on different roles of a person occurring as part of an organization. In this approach, each role is assigned to a set of permission to hold a place as a hierarchy among other entities. It includes two types of mappings, which are User Role Assignment (URA) and Role Permission Assignment (RPA) both is updated individually. Users can extend the URA without changing RPA by providing a predefined role to the new users. And also RPA assigns acceptable behaviors that are restricted to the user. The purpose of RBAC is that URA and PRA vary with less frequency than the permission of individual users. It has been adaptable for use in ubiquitous computing environments. In brief, RBAC includes the following:

- It lists the set of privileges related the user role.
- The normal action of the user is obtained.
- Privileges under role are controlled for allowance.
- If the user role is in the acceptable situation, then the user is authenticated.

5.3.4 Information Leakage

Because of the sensitive information particularly in expensive products, and concerning of users due to their information security, this matter is critical to solving. On the other hand, RFID systems only response with distinguished emitting signals to a query which is related to neighborhood domain. Leakage of information may occur without knowing to the users. Information leakage is more problematic by insiders when the property's value is high and if the leakage happens in information sharing and information accessibility the problem is more serious. Hence, it is important to develop security technology, which controls the leakage more and more at any time by enabling staff inside the company to access inside information at any time in any place supporting high work efficiency.

5.3.5 Cryptographic Protocols

With the proliferation of small and resource-constrained devices, lightweight cryptography has appeared as a new branch of cryptography which incurs lesser computational overhead. Protocols in ubiquitous computing mostly rely on three types of cryptographic algorithms asymmetric and asymmetric, while a majority of researchers focusing on a combination of both (hybrid approaches) to design light weight protocols.

5.3.6 Trust-Based Computation and Management

In a ubiquitous computing network, mobile nodes can engage themselves in a spontaneous interaction with other nodes. With the number of growing mobile devices participating as network nodes, it becomes a challenge to ensure its behavior properly. The importance of designing efficient trust models for such dynamic interaction and fast-changing network topologies has gained prominence over the past decade.

Trust models rely on the trust values that are assigned or derived based on the node's behavior. Some models assume direct interaction with the newly arrived node to assess its behavior, while other models use recommendations from previous nodes those interacted with the newcomer node to evaluate its behavior. If a node is predicted to be well-behaved, it can communicate with other nodes on the network. Malicious nodes are excluded from the network. It is important to ensure dynamics in trust computation by monitoring a node's actions and recomputing the assigned trust values since nodes may become non-cooperative over time.

5.3.7 Traceability

An opponent can record the transmitted message from a response to a target tag and establish a link between them. Through this link, the user's location information can be traced by an opponent. In the authentication situations the following must be considered:

- Read the RFID-tag from the product
- Transform of the RFID-tag for the database server
- Check for the authenticity of the RFID-tag
- Compare the RFID-tag with an organization that indicates authentication.

5.3.8 Biometrics

It implicates good properties to provide consistent and automated mechanisms for determining and confirming identity while, being less prominent. Finger print recognition, or face recognition techniques are faster than entering secure passwords and they do not require special devices like PDA. But, the biometric authentication techniques are heavily dependent on hardware in terms of accuracy and consistency. In any biometric technique, the principal concern is the biometric template and designing these elements represent unique personal information. Unfortunately, it is not possible to change or replace the biometric characteristics—they are an inherent part of the person unlike the other forms of authentication (such as secret knowledge or tokens, which can be simply changed if lost or stolen). Therefore, once lost or stolen, they are compromised and can no longer be used reliably. Even though, biometric authentication techniques still do not have a good and secure method of storing biometric properties, they preserve anonymity while providing enough flexibility to accommodate partial matches and reduce a suitable confidence level.

Conclusion: Security is one of the most prominent issues in ubiquitous computing. It enables communication devices to traverse openly, anywhere at any time. Therefore, the modern computing networks have become increasingly ubiquitous. When services are provided easily for all various networks and their users, the major concern is about the security. In this chapter we discussed the security issues in a ubiquitous computing environment such as Location privacy, vulnerability, attacks and we also explained some defenses concerning attacks.

References

1. Dragoni N, Massacci F, Walter T, Schaefer C (2009) What the heck is this application doing?—a security-by-contract architecture for pervasive services. Comput Secur 28(7):566–577
2. Pietro RD, Mancini LV (2003) \Security and privacy issues of handheld and wearable wireless devices. ACM Commun 46(9):74–79
3. Yau SS, Huang D, Gong H, Yao Y (2006) Support for situation awareness in trustworthy ubiquitous computing application software. J Softw Pract Eng 36(9):893–921
4. Weiser Mark (1991) The Computer for the 21st century. Sci Am 265(3):66–75
5. Weiser Mark (1994) The World is not a desktop. Interactions 1(1):7–8
6. Kang YB, Pisan Y (2006) A survey of major challenges and future directions for next generation pervasive computing. In: Proceedings of the 21st international symposium on computer and information sciences, pp 755–764
7. Poslad S (2009) Ubiquitous computing: smart devices, environments and interactions. Wiley-Blackwell
8. Varshney U, Vetter R (2002) Mobile commerce: framework, applications, and networking support. ACM/Kluwer Journal on Mobile Networks and Applications (MONET) 7(3):185–198
9. Spreitzer M, Theimer M (1993) Providing location information in a ubiquitous computing environment. In: Proceedings of SIGOPS '93, Dec 1993, pp 270–283
10. Garlan D, Siewiorek D, Smailagic A, Steenkiste P (2002) Project aura: towards distraction-free pervasive computing. IEEE Pervasive Comput 1:22–31
11. Beresford A, Stajano F (2003) Location privacy in pervasive computing. IEEE Pervasive Comput 2(1):46–55
12. Chaum DL (1981) Untraceable electronic mail, return addresses, and digital pseudonyms. Commun ACM 24(2):84–88, http://doi.acm.org/10.1145/358549.358563
13. Gruteser M, Grunwald D (2003) Anonymous usage of location-based services through spatial and temporal Cloaking. In: Proceedings of MobiSys 2003, the Usenix Association, San Francisco, CA, USA, pp 31–42
14. Jackson IW (1998) Who goes here? confidentiality of location through anonymity. PhD thesis, University of Cambridge
15. Danezis G, Lewis S, Anderson R (2005) How much is location privacy worth? In: Proceedings of Workshop on Economics of Information Security (WEIS), http://infosecon.net/workshop/pdf/locationprivacy.pdf
16. Spreitzer M, Theimer M (1993) Providing location information in a ubiquitous computing environment. In: Proceedings of SIGOPS '93, pp 270–283
17. Leonhardt U, Magee J (1998) Security considerations for a distributed location service. J Netw Syst Manage 6:51–70
18. Rannenberg K, Pfitzmann A, Muller G (1999) Multilateral security in communications, chapter IT security and multilateral security, Addison-Wesley-Longman, pp 21–29
19. Finkenzeller K (2000) RFID handbook: radio-frequency identification fundamentals and applications. Wiley
20. Romer K, Domnitcheva S (2002) Smart playing cards: a ubiquitous computing game. Journal for Personal and Ubiquitous Computing (PUC) 6
21. Wieselthier JE, Ephremides A, Michaels LA (1989) An exact analysis and performance evaluation of framed aloha with capture. IEEE Trans Commun 37(2):125–137
22. Sarma SE, Weis SA, Engels DW (2002) RFID systems and security and privacy implications. In: Workshop on Cryptographic Hardware and Embedded Systems, Lecture Notes in Computer Science, pp 454–470

23. Roberti M (2005) Understanding the EPC Gen 2 Protocol. RFID journal special report, Mar 2005
24. EPCglobal Inc. (2005) EPC™ Radio-Frequency Identity Protocols Class-1 Generation-2 UHF RFID, Jan 2005
25. Luo Z, Chan T, Li JS (2005) A lightweight mutual authentication protocol for RFID networks. In: IEEE international conference on e-business engineering, Oct 2005, pp 620–625
26. Garfinkel SL, Juels A, Pappu R (2005) RFID privacy: an overview of problems and proposed solutions. IEEE Secur Priv 3(3), pp 34–43, May–June 2005
27. Needham RM, Schroeder MD (1978) Using encryption for authentication in large networks of computers. Commun ACM 21(12):993–999
28. Norman DA (1998) The invisible computer: why good products can fail, the personal computer is so complex, and information appliances are the solution. MIT Press
29. Kohl J, Neuman C (1993) Thekerberos network authentication service (v5). RFC 1510, IETF, http://www.ietf.org/rfc/rfc1510.txt
30. Stajano F, Anderson R (1999) The resurrecting duckling: Security issues in ad-hoc wireless networks. In: Christianson B, Crispo B, Malcolm JA, Roe M (eds) Proceedings of the 7th international workshop security protocols, Lecture Notes in Computer Science, vol 1796, Springer, pp 172–182
31. Stajano F (2002) Security for ubiquitous computing. Wiley, http://www.cl.cam.ac.uk/~fms27/secubicomp/
32. Stajano F (2001) The resurrecting duckling—what next? In: Christianson B, Crispo B, Malcolm JA, Roe M (eds) Proceedings of the 7th international workshop security protocols, Lecture Notes in Computer Science, vol 2133, Springer-Verlag, pp 204–214
33. Anderson RJ, Kuhn MG (1996) Tamper resistance—a cautionary note. In: Proceedings of the second usenix workshop on electronic commerce, Usenix Association, Berkeley, pp 1–11
34. Sharif A, Khosravi M, Shah A (2013) Security attacks and solutions on ubiquitous computing networks. International Journal of Engineering and Innovative Technology (IJEIT) 3(4), Oct 2013, ISSN:2277-3754